Ventilation and Airflow in Buildings

BUILDINGS | ENERGY | SOLAR TECHNOLOGY

Ventilation and Airflow in Buildings

Methods for Diagnosis and Evaluation

Claude-Alain Roulet

Routledge
Taylor & Francis Group

LONDON AND NEW YORK

First published 2008 by Earthscan

2 Park Square, Milton Park, Abingdon, Oxon OX14 4RN
711 Third Avenue, New York, NY 10017,USA

Routledge is an imprint of the Taylor & Francis Group, an informa business

Firstissuedinpaperback2016

Typesetting by 4word Ltd, Bristol
Cover design by Paul Cooper

A catalogue record for this book is available from the British Library

Library of Congress Cataloging-in-Publication Data

Roulet, Claude-Alain.
Ventilation and airflow in buildings : methods for diagnosis and evaluation / Claude-Alain Roulet.
      p.   cm.
   Includes bibliographical references and index.
     ISBN-13: 978-1-84407-451-8 (hardback)
     ISBN-10: 1-84407-451-X (hardback)
   1. Ventilation–Handbooks, manuals, etc. 2. Air flow–Measurement–Handbooks, manuals, etc. I. Title.
   TH7656.R68 2008
   697.9′2–dc22

                                                                                    2007034790

ISBN-13: 978-1-84407-451-8 (hbk)
ISBN-13: 978-1-138-98669-5 (pbk)

Contents

List of Figures and Tables

Figures

Tables

Preamble

This book includes information already published by the author in scientific journals and in an Air Infiltration and Ventilation Centre (AIVC) technical note (Roulet and Vandaele, 1991), now sold out. Part of the content of Chapters 2, 3 and 5 was also published by the author in a book edited by H. Awbi (Awbi, 2007).

Roulet, C.-A. and L. Vandaele, 1991, Airflow patterns within buildings: Measurement techniques. *AIVC Technical Note 34*, AIVC, Bracknell, 265pp, order at inive@bbri.be

Awbi, H., 2007, *Ventilation Systems, Design and Performance*, Taylor and Francis, London, 522pp

Introduction

Why ventilate?

Without ventilation, a building's occupants will initially be troubled by odours and other possible contaminants and heat. Humidity may rise because of indoor moisture sources such as the occupants, laundry, cooking and plants; thus enhancing moisture hazards (for example, mould growth and condensation). Oxygen will nevertheless not be missed until much later. The purpose of ventilation is to eliminate airborne contaminants, which are generated both by human activity and by the building itself. These are:

- bad odours, to which people entering the room are very sensitive;
- moisture, which increases the risk of mould growth;
- carbon dioxide (CO_2) gas, which may induce lethargy at high concentrations;
- dust, aerosols and toxic gases resulting from human activity, as well as from the building materials (in principle, 'clean' materials should be chosen for internal use, but this is not always possible);
- excessive heat.

The airflow rate required to ensure good indoor air quality depends upon the contaminant sources' strengths and on their maximum acceptable concentration: the larger the contaminant sources' strengths or the smaller the maximum acceptable concentration, the greater the required ventilation rate is.

During the heating season in well-designed and clean buildings, the occupants are the main source of contaminants (mostly odours and water vapour). The airflow rate should then be between 22 cubic metres per hour (m^3/h) per person, which limits the CO_2 concentration to about 1000 parts per million (ppm) above the outdoor concentration, and 54 m^3/h per person, which limits the CO_2 concentration to about 400 ppm above the outdoor concentration – meaning that less than 10 per cent of people entering the room will be dissatisfied by the odour (CEN, 2006). Airflow rates should be much greater in poorly insulated buildings (where there is a risk of mould growth and water vapour condensation), or in spaces where there is a particular source of contamination, including spaces where smoking is allowed.

In summer, the minimum airflow rate may be much greater than the hygienic airflow rate in order to evacuate heat or provide cooling draughts.

xiv   *Ventilation and Airflow in Buildings*

However, when the outdoor temperature exceeds indoor temperatures, it may be wise to reduce the ventilation rate, only allowing high levels of ventilation at night when the outdoor temperature is low.

Ventilation is hence not only essential to ensure an acceptable indoor air quality, but is also often used to improve thermal comfort. For this air heating or cooling, air conditioning (including air humidity control) or free cooling (increasing the outdoor airflow rate to cool down the building fabric) are used. In order to achieve these goals, several conditions should be met:

- Airflow rates should be adapted to need: if too low, good air quality will not be achieved, or draughts, noise and energy waste may result from an excessive airing.
- The air should be well distributed: ideally, the fresh air should reach any occupied zones first and contaminated air should be quickly extracted.
- The air supply should not decrease comfort. It should not cause complaints about draughts, noise or poor air quality.
- The air supplied by ventilation systems should be clean and, where appropriate, should comply with the temperature and moisture requirements.

In addition, to comply with a sustainable development policy, the ventilation systems should be energy efficient and should perform as required using a minimum amount of energy.

Why assess airflows in buildings?

The conditions listed above are most likely to be met when the building and its ventilation system are not only well designed and built, but also well commissioned. Commissioning a ventilation system involves carrying out measurements to check that it performs as expected. When these conditions are not met or when there are problems, measurements may help in finding the causes of the problem and in fixing them.

In order to show the usefulness of measurements, some results from investigations performed on several air handling units are shown below. It should be emphasized that these ventilation units were not selected because they had problems. The air handling units, located in different buildings, were measured in several measurement sessions (Roulet *et al.*, 1999). In some units, the airflow rate was far from the design values, or there was unexpected recirculation.

Outdoor airflow rate

The comparison of design and measured outdoor airflow rate per person in 12 buildings is shown in Figure 0.1. It can be seen that in several buildings the airflow rate per person is larger than $50\,\mathrm{m^3/h}$, and surpasses $200\,\mathrm{m^3/h}$.

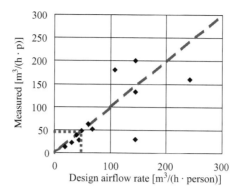

Figure 0.1 *Design and measured outdoor airflow rate per person in
12 buildings*
Source: Roulet *et al.*, 1999.

Measured airflow rates differ from the design values in many ventilation systems. Figure 0.2 shows the relative differences between measured and design outdoor airflow rate in 37 air handling units, i.e.:

$$\frac{\text{Measured airflow rate} - \text{design airflow rate}}{\text{Design airflow rate}}$$

These biases range from −79 per cent to +67 per cent. Only 11 units are within the ±10 per cent range.

Recirculation rates

Some recirculation of air is often planned to distribute heat or cold without conditioning too much outdoor air. This, however, decreases the global indoor air quality, since the contaminants generated within the building are recirculated throughout the whole building. Therefore, recirculation may not be desirable. In any case, recirculation should be controlled. Design and

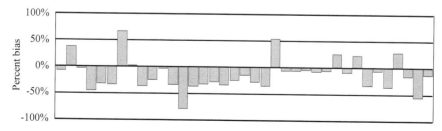

Figure 0.2 *Relative difference between measured and design outdoor
airflow rate in 37 air handling units*
Source: Roulet *et al.*, 1999.

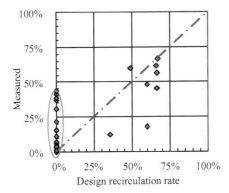

Figure 0.3 *Comparison of design and measured recirculation rate in 34 air handling units*

Source: Roulet *et al.*, 1999.

measured recirculation rates are compared in Figure 0.3. These are seldom the same. Even worse: as shown in Figure 0.4, out of 27 units planned without recirculation, 30 per cent have shown a recirculation rate of more than 20 per cent.

Exfiltration

In supply and exhaust units, both airflow rates are either balanced, or the supply flow rate is increased a little to make the building slightly overpressured. When the envelope is not airtight, and when the balance between supply and exhaust air is too large, air leaks through the envelope. This does not have much influence on indoor air quality, but may strongly decrease the efficiency of the heat recovery. In some buildings, as much as 100 per cent of the supply air is lost in this way, as illustrated in Figure 0.5, which compares

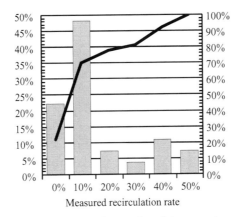

Figure 0.4 *Histogram and cumulated frequencies of measured recirculation rates in air handling units designed without recirculation*

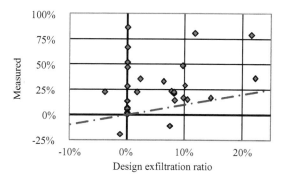

Figure 0.5 *Design and measured exfiltration ratios compared in 30 units*
Source: Roulet *et al.*, 2001a.

the design and measured exfiltration ratios, i.e. parts of the supply air leaking through the building envelope in 30 units.

Ventilation efficiency

An efficient airing supplies the occupants with fresh air and blows polluted, old air in unoccupied spaces. The ventilation efficiencies in several rooms, assessed using the method described in Chapter 3, are illustrated in Figure 0.6.

Rather high ventilation efficiency, indicating piston-type ventilation, can be seen in rooms 6 to 10, which are high auditoriums. The normal-height rooms 2 to 5 present a complete mixing, while room 1 shows poor ventilation efficiency, partly explained by the fact that supply and exhaust are both located at the ceiling in this room.

These few examples clearly illustrate the usefulness of measurements to detect dysfunction.

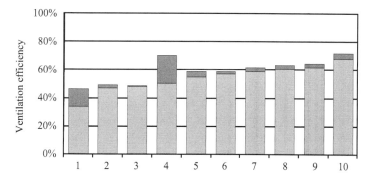

Figure 0.6 *Ventilation efficiency in some ventilated areas*
Note: Dark bands are uncertainty bars. Note that in one unit the efficiency is below 50 per cent, indicating shortcuts and dead zones.
Source: Figure drawn from a table published in Roulet *et al.*, 2001a.

When assess airflows in buildings?

Ventilation performance should be checked early to detect potential problems and to optimize the overall performance of the ventilation system. This includes:

- appropriate airflow rates;
- negligible leakage and shortcuts;
- high ventilation efficiency;
- high fan efficiency;
- clean air and so on.

This check should be performed:

- when commissioning the ventilation system in order to control that the system is built according to the specifications;
- if there are indoor air quality problems to help in finding the causes;
- before refurbishing the system in order to accurately know where there are potential problems, which should be cured by the refurbishing.

Available methods to assess airflow rates and related quantities

This section briefly presents the methods described in the book in order to guide the user in the choice of the appropriate method. The section also proposes adapted methods for different purposes and gives a general guideline for planning measurements.

Airflow rates in buildings and in handling units

Chapter 1 describes in detail the method of using tracer gases for assessing airflow rates between indoor and outdoor spaces, and between indoor spaces. Such measurements may be useful to check if the ventilation is sufficient. The method also allows checks to assess if the airflows follow defined paths from room to room, as required in some buildings, such as laboratories handling dangerous substances.

Similar methods, described in Chapter 2, test if airflow rates in air handling units correspond to the design values, and detect possible leakage or parasitic airflows in such units. Such measurements are also useful to check the power efficiency of fans (see Chapter 5, 'Energy for ventilation') and energy efficiency of heat exchangers (Chapter 5, 'Heat exchange efficiency').

Age of air and ventilation efficiency

The longer the air stays at a given location, the larger will be its concentrations of various contaminants. The age of the air, i.e. the time spent in the building since the outdoor air entered it, can be measured using tracer gases. The effectiveness of the ventilation in appropriately distributing the air in the ventilated space or in evacuating contaminants emitted at a given location can be assessed using the methods described in Chapter 3.

When applied in the air handling unit, such measurements can be performed simultaneously with measurements of airflow rates, thus reducing the amount of work required. In a single measurement campaign, the mean age of the air in the ventilated space, the efficiency of the ventilation system, the supply, exhaust and recirculation flow rates can be measured as well as the air leakage in both directions through the building envelope.

Airtightness

The building envelope should be reasonably airtight to ensure the efficiency of a mechanical ventilation system, and this system itself should have airtight ducts in order to distribute the air appropriately throughout the ventilated space. Chapter 4 describes the measurement methods. The general methods are described first and their application to the building envelope can be found in Chapter 4, 'Airtightness of buildings', while the application to air ducts or ductworks are in Chapter 4, 'Measurement of airtightness of a duct or network'.

Energy efficiency

To comply with a sustainable development policy and also to reduce the emission of greenhouse gases, energy efficiency of any system should be improved. This can be achieved without the use of systems powered by non-renewable energy. In our case, a mechanical ventilation system could be applied to an appropriate building design, allowing natural ventilation. However, this is not always possible, and in some cases – for example in cold or hot countries where heat recovery on indoor air actually allows energy savings – may even be counterproductive.

Therefore, the mechanical ventilation systems should be designed, built, commissioned and maintained with the aim of ensuring good indoor air quality at reduced energy use. Chapter 5 proposes various methods, checklists and propositions to measure, check and improve the energy efficiency of ventilation systems and components. They include:

- efficiency of heat recovery ('Heat exchange efficiency');
- effectiveness of heat exchangers('Heat exchangers');
- fan power efficiency ('Energy for ventilation');
- energy effects of indoor air quality measures ('Energy effects of IAQ measures').

Contaminants in air handling units

Unfortunately, in practice major sources of indoor air contaminants are components in air handling units and ventilation systems (Bluyssen *et al.*,

1995, 2003). This can, however, be avoided by appropriate design and maintenance. Chapter 5, 'Energy effect of IAQ measures', lists the sources and causes of pollution in ventilation systems, proposes measurement protocols, and sets out maintenance procedures and strategies to improve the quality of the air delivered by mechanical ventilation systems.

Common techniques

The description of techniques and methods used for the measurements and their interpretation is detailed in Chapter 7. This chapter includes:

- the general description of tracer gas dilution techniques;
- ways of expressing concentrations and flow rates;
- mathematical identification methods;
- error analysis.

Airflow Rates in Buildings

This chapter intends to help the reader to measure airflow rates and air change rates in buildings and rooms, independently of a mechanical ventilation system. It presents the techniques used to measure the airflow rate entering the measured zone (single-zone measurements) and to measure inter-zone airflows (multi-zone measurements).

A building zone or a zone is a space that can be considered as homogeneous from the point of view of air quality or, more technically, a space in which each tracer gas is homogeneously distributed. In practice, it is a room or a set of adjacent rooms that have much larger airflow rates between them than to or from other zones or the outdoor space.

The measurement techniques presented here are all based on the use of tracer gases that are injected into the air and analysed in air samples after mixing. More detailed information on the tracer gases themselves, on appropriate injection and sampling methods and on tracer gas analysers is given in Chapter 7, 'Tracer gas dilution techniques'.

Single-zone measurements

The tracer gas is injected in a space, mixed into the air and its concentration is measured. Various strategies can be used for assessing the airflow rate entering the space: recording and interpreting the concentration decay after having stopped the injection, monitoring the tracer gas concentration when injecting the gas at constant rate, or measuring the tracer gas flow rate required for keeping its concentration constant. Airflow rates are obtained by interpreting the evolution with time of either the tracer gas concentration or the injection rate. The interpretation methods are based on the mass conservation of tracer gas and of the air.

Mass conservation of tracer gas and air

The tracer gas injected in a building space is uniformly mixed into the air. The conservation of the mass of tracer gas within a single zone in contact with the outdoor environment is then:

$$\frac{dm}{dt} = I + C_o Q_{oi} - C_i Q_{io} \tag{1.1}$$

where:

m is the mass of tracer gas in the zone (kg);
I is the injection rate of the tracer gas (kg/s);
C is the tracer gas mass concentration;
Q is the mass airflow rate (for example, Q_{io} is airflow rate from indoor to outdoors);
i subscript for internal environment;
o subscript for external environment.

In addition, the conservation of the mass of air gives:

$$Q_{oi} = Q_{io} \tag{1.2}$$

The mass of tracer in the zone is related to the mass of air M by:

$$m = C_i M \tag{1.3}$$

where C_i is the concentration of the tracer gas in the indoor air. Combining the last three equations, we get:

$$M\frac{dC_i}{dt} = I + Q_{io}(C_i - C_o) \tag{1.4}$$

since M is very close to a constant if the temperature is constant. In principle, this equation can directly provide the airflow rate:

$$Q_{io} = \frac{I - M\dfrac{dC_i}{dt}}{\Delta C} \tag{1.5}$$

writing $\Delta C = C_i - C_o$.

This method is, however, very inaccurate, since very quickly the concentration may vary at random because of turbulence and non-homogeneities. It is therefore better to take a time average by integrating it for a given period of time:

$$\int_t^{t+\Delta t} Q_{io}\, dt = \int_t^{t+\Delta t} \frac{I}{\Delta C}\, dt - M\int_t^{t+\Delta t} \frac{dC_i}{\Delta C} \tag{1.6}$$

hence:

$$\int_t^{t+\Delta t} Q_{io}\, dt = \int_t^{t+\Delta t} \frac{I}{\Delta C}\, dt - M[\ln(\Delta C(t)) - \ln(\Delta C(t + \Delta t))] \tag{1.7}$$

or, dividing both members by Δt

$$\langle Q_{io}\rangle = \left\langle \frac{I}{\Delta C}\right\rangle - \frac{M}{\Delta t}\ln\left(\frac{\Delta C(t)}{\Delta C(t + \Delta t)}\right) \tag{1.8}$$

where the quantity between brackets $\langle\rangle$ is averaged over the time period Δt.

This solution can be simplified, depending on the way the tracer is injected.

Tracer decay, no injection

A suitable quantity of tracer gas is injected to achieve a measurable initial concentration $C_{i,0}$. At time t_0, this injection is stopped and $I = 0$ afterwards. From Equation 1.4, it can be found that the concentration decays with time according to:

$$C = C(t_0) \exp\left(-\frac{Q_{io}}{M} t\right) \tag{1.9}$$

The quantity

$$\tau_n = \frac{M}{Q_{io}} \tag{1.10}$$

is called the nominal time constant of the measured zone. It is the ratio of the mass of air contained in the zone to the mass airflow rate. It is also the time needed to introduce a mass of new air equal to that contained in the zone.

Since $I = 0$, Equation 1.8 becomes:

$$\langle Q_{io} \rangle = -\frac{M}{\Delta t} \ln\left(\frac{\Delta C(t)}{\Delta C(t + \Delta t)}\right) \tag{1.11}$$

This equation allows easy calculation of the airflow rate from the measurement of concentration at two instants. This method is called the decay method. It is a direct measurement of the nominal time constant, and also provides an unbiased estimate of the mean airflow rate.

Constant injection rate

If the injection rate is constant, the solution of Equation 1.4 is:

$$\Delta C = \Delta C(t_0)\frac{I}{Q_{io}}\exp\left(-\frac{t}{\tau_n}\right) + \frac{I}{Q_{io}} \tag{1.12}$$

Using identification technique (see Chapter 7 'Identification methods'), both τ_n and Q_{io} (hence also M) can be obtained. This method is, however, of easy use only when Q_{io} is constant. In this case, the exponential term becomes negligible after three or more time constants, and

$$\Delta C = \frac{I}{Q_{io}} \tag{1.13}$$

or

$$Q_{io} = \frac{I}{\Delta C} \tag{1.14}$$

The result is biased (underestimated) if the airflow rate is not constant.

Constant concentration

Using an electronic mass flow controller monitored by the tracer gas analyser, the concentration of tracer gas can be maintained constant by varying the injection rate in an appropriate way. In this case, the time derivative of the concentration is zero and Equation 1.4 becomes very simple:

$$I + Q_{io}(C_i - C_o) = 0 \qquad (1.15)$$

hence:

$$Q_{io} = \frac{I}{(C_i - C_o)} = \frac{I}{\Delta C} \qquad (1.16)$$

This method provides an unbiased estimate of the airflow rate, even when it varies with time.

Pulse injection

The method can also be used with the tracer injected as a short pulse at time t_0. The injected mass M will result in a tracer gas concentration at sampling location that varies with time, starting from background concentration C_0, growing and then decaying back to background concentration. Let $C(t)$ be the tracer concentration above background. The total mass of tracer gas passing at the sampling location is then:

$$M = \int_{t_0}^{\infty} C(t)Q(t)\,dt \qquad (1.17)$$

An approximation to infinite time can be good enough when the experiment (and the integral) is stopped at time t_f, when the concentration is close enough to background.

Since both functions $C(t)$ and $Q(t)$ are positive and continuously derivable, we can apply the integral mean value theorem (Axley and Persily, 1988), that is:

$$M = Q(\tau) \int_{t_0}^{t_f} C(t)\,dt \quad \text{with} \quad t_0 < \tau < t_f \qquad (1.18)$$

This means that there exists a time τ during the experiment, when the airflow rate Q has a value satisfying the above equation. The knowledge of the injected mass M and measurements of the integral of the concentration downwind can then provide a value of Q.

The integral of the concentration can be calculated from the mass of tracer m sampled downwind from the injection port by pumping the air at known rate $Q_{s,i}$ through a tube. The sampled mass is related to concentration by:

$$m = \int_0^{\infty} C(t)Q_s(t)\,dt \qquad (1.19)$$

Applying again the integral mean value theorem, we get:

$$m = Q_{s,i}(\tau') \int_0^{t_f} C(t)\,dt \quad \text{with } 0 < \tau' < t_f \tag{1.20}$$

Combining Equations 1.18 and 1.20, we get:

$$Q_i(\tau) = Q_s(\tau')\frac{m}{M} \tag{1.21}$$

There are sampling pumps with controlled constant flow rate. These easily enable $Q_{s,i}$ to be kept constant, and in most cases it is possible to keep the airflow rate constant in supply and exhaust during the experiment.

Simple and cheap air change rate measurement using CO_2 concentration decays

Method

CO_2 generated by occupants can be used as a tracer gas, since it is easy and cheap to measure. There are compact and light CO_2 analysers on the market that include a data logger. Peak value of the CO_2 concentration during occupancy is an indicator of the minimum airflow rate per person. Analysis of the decays observed when the occupants leave the building provides the nominal time constant of the ventilated space, which is directly dependent on the outdoor airflow rate from the ventilation system and infiltration. Depending on the state of the ventilation system during the decay, this method provides either the total outdoor airflow rate provided by the system, or the infiltration rate. When combined with a simple pressure differential measurement, this method can also be used to check airtightness of building envelopes.

Equivalent outdoor airflow rate

Air may enter into a measured zone not only directly from outdoors, but also from neighbouring zones, whose CO_2 concentration may differ from outdoor air. These inter-zone airflows influence the CO_2 concentration in the measured zone, but can be measured only with complex and expensive techniques (see Chapter 1, 'Application to buildings, multi-zone'). The concept of equivalent outdoor airflow rate is introduced to offset this inconvenience. It corresponds to the outdoor airflow rate that would result in the same CO_2 concentration in the measured room without inter-zone airflows. In the following, this quantity is referred to as 'outdoor airflow rate'.

Equivalent outdoor airflow rate per person

An adult person produces on average and for most of the time (i.e. when quiet or doing light work with about a 100 W metabolic rate) about 20 litres per hour (l/h) of CO_2. At steady state, and assuming that occupants are the only CO_2 sources, the equivalent outdoor airflow rate per person, Q_e, is related to CO_2

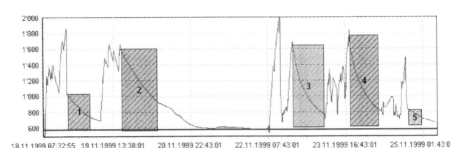

Figure 1.1 *Records of CO_2 concentration in an office room*
Source: Roulet and Foradini, 2002.

concentration C (C_i indoors and C_o outdoors) by:

$$Q_e = \frac{S}{C_i - C_o} \tag{1.22}$$

where S is the CO_2 source strength, i.e. about 20 l/h. The equivalent outdoor airflow rate per person can then be assessed during the periods of time when steady state can reasonably be assumed, that is when the CO_2 concentration is constant.

Example of application

CO_2 concentration was recorded every five minutes during several winter days in an office room occupied by one person.

The evolution of CO_2 concentration is shown in Figure 1.1. A base outdoor concentration of about 600 ppm was determined from the minimum values at the end of long decay periods (weekends). This base concentration is deducted from the CO_2 concentration to get the increase resulting from indoor sources.

On 19 November, a CO_2 concentration of about 1500 ppm is observed. This corresponds to an equivalent outdoor airflow rate at $22 \, m^3/(h \cdot person)$, obtained by natural ventilation. Decay periods are selected in the record (rectangles in Figure 1.1). They correspond to night or weekend periods, without occupancy, when windows and doors are closed and the air change rate results from infiltration only. The average nominal time constant from these five decays is found to be 10 ± 2 hours.

Application to buildings, multi-zone

Most buildings include several interconnected zones. In order to measure not only the airflow between internal and external environments but also inter-zone flows, either several tracer gases should be used simultaneously (injecting each of them in a different zone), or several experiments should be conducted successively, injecting the tracer successively in the different zones, and assuming that the measurement conditions, in particular the airflow pattern, do not change during the measurement campaign. This section describes ways of interpreting the records of tracer gas injection rates and concentration in the different zones to get the airflow rates between zones, as well as airflow to and from outdoors. For more information on tracer gases and analysers, see Chapter 7, 'Tracer gas dilution techniques'.

Let us assume that there are N zones in the measured building, denoted by the suffixes i and j, into which, in principle, N different tracers, denoted by the index k, are injected. In principle, each zone receives only one tracer, but the equations presented below allow the use of several gases in the same zone. No tracer is injected in the outside air (zone 0), which is assumed to be of infinite volume. However, the tracer concentration in that zone may differ from zero.

The multi-zone tracer gas theory is based on the conservation of the mass of tracer gas and of air and on the following three assumptions:

1   In each zone, tracer concentrations are always homogeneous.
2   The atmospheric pressure is constant.
3   The injection of tracer gas does not change the density of air.

The first assumption is the weakest. In practice, homogeneous concentration may only be achieved by the use of mixing fans, but these fans may affect infiltration conditions.

The other two hypotheses are easily satisfied because the short-time relative variations of atmospheric pressure are of the order of 0.01 per cent (daily variations of the order of a per cent) and tracer gases are generally injected at relatively low concentrations (10^{-4} or less).

Conservation of the masses of air and tracer gas *k* in zone *i*

In each zone, the rate of change of the air mass m_i equals the sum of the incoming flows minus the sum of the outgoing flows:

$$\frac{dm_i}{dt} = \sum_{j=0}^{N} Q_{ij}(1 - \delta_{ij}) - \sum_{j=0}^{N} Q_{ji}(1 - \delta_{ij}) \tag{1.23}$$

| Change in mass | Incoming flow rates | Outgoing flow rates |

where δ_{ij} is equal to one only when $i = j$, and zero otherwise. The sum is then over all terms for which $i \neq j$. Note that, in most cases, the left-hand side of these equations is close to zero and can be neglected.

The conservation equation of the mass of tracer, k, in the zone, i, states that the change of tracer mass within the zone is the sum of the mass of injected tracer and the mass of tracer contained in the air entering the zone, minus the mass of tracer contained in the outgoing air:

$$\frac{dm_{ik}}{dt} = I_{ik} + \sum_{j=0}^{N} C_{jk}Q_{ij}(1 - \delta_{ij}) - C_{ik}\sum_{j=0}^{N} Q_{ji}(1 - \delta_{ij}) \tag{1.24}$$

| Variation | Injection | Inflow | Outflow |

where:

m_{ik} is the mass of tracer gas k in zone i;
I_{ik} is the injection rate of tracer gas k in (or just upwind of) zone i;
C_{jk} is the concentration of tracer k in zone j;
C_{ik} is the concentration of tracer gas k in zone i;
Q_{ij} is the airflow rate from node j to node i.

An extension of assumption 1 above is implicit in this equation, that is:

4   The airflow entering a zone does not modify the homogeneity of the concentration of tracer gases in that zone, i.e., an immediate and perfect mixing is assumed.

If there are N tracers or N different sets of measurements using a single tracer injected at various rates in the various zones, Equations 1.23 and 1.24 above give a full set of $N(N+1)$ equations. Therefore, this allows the $N(N+1)$ flows between all the zones, including the outdoor air as the zone zero to be determined. There are two methods to transform this set of equations before solving. Since they each have various advantages and disadvantages, they are both described below.

Global system of equations

The most common technique to be found in the literature (Sinden, 1978; Sherman *et al.*, 1980; Perera, 1982; Sandberg, 1984) is the following.
    Let us express by Q_{ii} the sum of all the flows entering the zone i:

$$Q_{ii} = \sum_{j=0}^{N} Q_{ij}(1 - \delta_{ij}) \tag{1.25}$$

Using the above notation and taking apart the flows coming from outside, Equation 1.24 becomes:

$$M_i \frac{dC_{ik}}{dt} = I_{ik} + \sum_{j=1}^{N} C_{kj}Q_{ij}(1 - \delta_{ij}) + C_{0k}Q_{i0} - C_{ik}Q_{ii} \tag{1.26}$$

Since any change in the outdoor level of tracer gas concentrations, C_{0k}, will be negligible, these levels are the base levels of tracer gas concentrations anywhere else. In this case the tracer mass balances expressed in Equation 1.26 can be written in a matrix form:

$$\frac{d}{dt}[\boldsymbol{M} \cdot \boldsymbol{C}] + \boldsymbol{Q} \cdot \boldsymbol{C} = \boldsymbol{I} \tag{1.27}$$

where each row of the $N \times N$ matrices $\boldsymbol{M} \cdot \boldsymbol{C}$, $\boldsymbol{Q} \cdot \boldsymbol{C}$ and $\boldsymbol{I}$ corresponds to a zone and each column to a given tracer gas. More specifically:

$\boldsymbol{M}$ is a diagonal matrix whose elements are the masses of air contained in each zone:

$$m_i = \rho_i V_i \text{ or } \boldsymbol{M} = \rho V$$

where ρ is the diagonal matrix of the air densities in the zones, ρ_i, and $\boldsymbol{V}$ the diagonal matrix of the volumes of the zones, V_i.

C contains the differences in mass concentrations $C_{ik} - C_{0k}$ of gas k in zone i.

I is the matrix containing the mass flow rates I_{ik} of the tracer, k, in zone i. In usual measurements, this matrix is diagonal.

Q is the so-called flow matrix containing, the off-diagonal elements $(j \neq i)$ being $-Q_{ij}$, where Q_{ij} represents the mass flow rates from zone j to zone i. The diagonal elements with $j = i$ contain the sum of the flows leaving the zone i, as defined in Equation 1.25.

In Equation 1.26, i and j run from 1 to N and this system results in N^2 equations for the inter-zonal flows. The mass flows to and from outside are given by Equation 1.23.

When steady state is reached for tracer gas concentration, Equation 1.27 becomes:

$$\boldsymbol{Q} \cdot \boldsymbol{C} = \boldsymbol{I} \qquad \text{hence} \qquad \boldsymbol{Q} = \boldsymbol{I} \cdot \boldsymbol{C}^{-1} \tag{1.28}$$

This method looks very attractive, but has several disadvantages. First, the number of tracer gases is limited, and it is therefore often impossible or at least very impractical to have tracer gases injected in each zone. In this case, the C-matrix is not square and cannot be inverted. In addition, this method may give non-zero values to non-existent airflow rates, or even provide negative airflow rates. For this reason, the node-by-node method, which allows writing equations containing only significant airflow rates, was developed.

Zone by zone systems of equations

Another presentation of the same model is found in Roulet and Compagnon (1989). It is obtained as follows.

Combining Equations 1.23 and 1.24, then taking into account that $m_{ik} = m_i \cdot C_{ik}$, and using:

$$\frac{dm_{ik}}{dt} = m_i \frac{dC_{ik}}{dt} + C_{ik} \frac{dm_i}{dt} \tag{1.29}$$

we finally get:

$$m_i \frac{dC_{ik}}{dt} = I_{ik} + \sum_{j=0}^{N} (C_{jk} - C_{ik}) Q_{ij} (1 - \delta_{ij}) \tag{1.30}$$

For each zone i, these N equations give the N flows, Q_{ij} $(j = 0, \ldots, i-1, i+1, \ldots, N)$. The flows, Q_{ji}, are obtained from the same equations applied to zone j and the remaining flows, Q_{0i}, are given by Equations 1.23.

Further interpretation of the flow matrix

The final result of the measurements is the flow matrix $\boldsymbol{Q}$ defined above. Further information can be deduced from this flow matrix, as shown in the following discussion (Sandberg, 1984).

Properties of the flow matrix

The total outdoor airflow rate to each zone, i, is easily obtained by summing the columns of the flow matrix:

$$Q_{i0} = \sum_{j=1}^{N} Q_{ij} \tag{1.31}$$

And the total exfiltration airflow rate from each zone, i, is the sum of the lines of the flow matrix:

$$Q_{0i} = \sum_{i=1}^{N} Q_{ij} \tag{1.32}$$

If there is no totally isolated chamber in the measured system, and if there is some air exchange with outside (as is the case with any usual building), the flow matrix determinant, $|\mathbf{Q}|$, is positive and $\mathbf{Q}$ has an inverse, $\mathbf{Q}^{-1}$.

The elements of this inverse $\mathbf{Q}^{-1}$ are given by:

$$W_{ji} = \frac{A_{ji}}{|\mathbf{Q}|} \tag{1.33}$$

where A_{ji} are the cofactors of the element Q_{ij} in $\mathbf{Q}$.

Transfer of contaminants between zones

The basic equations applied to the case where a constant flow rate, I_{ik}, of a contaminant, k, is applied in each zone, i, leads to an equilibrium concentration (for constant airflow rates) that is:

$$\mathbf{C}(\infty) = \mathbf{Q}^{-1}\mathbf{I} \tag{1.34}$$

It follows that the equilibrium concentration in room, j, resulting from a contaminant, k, released only in room, i, is:

$$C_{jk}(\infty) = W_{ji}I_{ik} \tag{1.35}$$

and the non-diagonal elements of $\mathbf{Q}^{-1}$ are hence the transfer indexes defined in Sandberg (1984).

Using a simple inversion of the flow matrix, much information on the possible spreading of contaminants can be obtained.

Age matrix and mean age of air

The τ matrix is defined as:

$$\tau = \mathbf{Q}^{-1}\mathbf{M} \tag{1.36}$$

or, under the assumptions of constant, uniform temperature:

$$\tau = \mathbf{q}^{-1}\mathbf{V} \tag{1.37}$$

Where $\mathbf{q}$ is the volume flow matrix and $\mathbf{V}$ a diagonal matrix with the volumes V_{ii} of room i on the diagonal. It is shown (Sandberg, 1984) that the row sums of the τ matrix are the mean age of air in the corresponding rooms:

$$\langle \tau_i \rangle = -\sum_{j=1}^{N} \tau_{ij} \tag{1.38}$$

This relation enables the measurement of the room mean age of air to be made, even in rooms where there are several outlets or several ways for the air to leave the room.

Equations for volume flow rates

All equations above are based on mass balance, and hence include mass airflow rates and mass concentrations. However, for practical reasons, volume flow rates and volume concentrations are of common use. Therefore, the basic equations should be adapted as shown below.

The mass of the tracer k in the zone i is:

$$m_{ik} = \rho_{ik} V_{ik} = \frac{\rho_i V_i C_{ik}}{1 - C_{ik}} \cong \rho_i V_i C_{ik} = \rho_{ik} V_i c_{ik} \tag{1.39}$$

since $C_{ik} \ll 1$.

The tracer density is defined by $\rho_{ik} = m_{ik}/V_{ik}$ where the volume, V_{ik}, is defined at atmospheric pressure, p. Using the perfect gas law for tracer k:

$$pV_{ik} = RT_i \frac{m_{ik}}{M_k} \tag{1.40}$$

where R is the molar gas constant, $R = 8313.96$ [J/(K·kmole)], M_k the molar mass of the tracer, k, and T_i is the absolute temperature of zone i. The density of tracer k in zone i can be computed:

$$\rho_{ik} = \frac{pM_k}{RT_i} \tag{1.41}$$

This is also valid for the density of air, by simply omitting the suffix k and using the average molecular weight ($M \cong 29$ g/mole) of the air.

Introducing this in Equations 1.23 and 1.26 gives the set of balance equations to be used when handling volumes instead of masses. Equation 1.30 becomes:

$$\frac{V_i}{T_i} \frac{dc_{ik}}{dt} = \frac{i_{ik}}{T_k} + \sum_{j=0}^{N} \frac{(c_{jk} - c_{ik})}{T_j} q_{ij}(1 - \delta_{ij}) \tag{1.42}$$

where:

T is the absolute temperature of zone i or j, or of tracer k, depending on the subscript;

c_{ik} is the volume concentration of tracer k in zone i;

i_{ik} is the volume injection rate of tracer k in zone i;

q_{ij} is the volume flow rate from zone j to zone i.

The air mass conservation (Equation 1.23) is rewritten as:

$$q_{0i} = T_i \sum_{j=0}^{N} \frac{q_{ij}(1 - \delta_{ij})}{T_j} - \sum_{j=1}^{N} q_{ji}(1 - \delta_{ij}) + \frac{V_i}{T_i} \frac{dT_i}{dt} \qquad (1.43)$$

These last two systems include $N + 1$ equations for $N + 1$ unknowns, q_{ij}, for each zone i.

It should be noticed that Equation 1.43 can be simplified, and becomes similar to Equation 1.23 if indoor and outdoor temperatures are close to each other and if the internal temperature is constant. This means that, provided such conditions are realized, the volume conservation equation can be used instead of mass conservation.

Summary of the various tracer gas methods

The different tracer gas techniques can be broadly divided into two categories: steady-state methods, which directly measure the flow rate, Q, and transient methods, which measure the nominal time constant, τ_n, or the air change rate, n. The steady-state techniques are based on recording steady-state concentrations or concentrations integrated over a long time, while transient methods are based on recording the change in tracer gas concentration. The different tracer gas techniques and their properties are given in Tables 1.1 and 1.2.

If airflow varies with time, only the two-point decay and the constant concentration methods give a correct estimate of the average flow. The constant injection method underestimates the average flow rate if the integration time is much longer than the period of flow variation.

Table 1.3 gives a summary of multi-zone methods. As far as single-zone measurements are concerned, the following conclusions can be stated:

- It appears that decay, pulse and step-up methods require the least measurement time and usually the least preparation. However, with the exception of the two-point decay method, they give a biased estimate of a variable air change rate. These biases remain small if the measurement period is limited to times close to the nominal time constant.

Table 1.1 *Summary of different injection strategies*

Tracer injection strategy	Direct result	Cost
Pulse injection	$Q^{\dagger}$	Moderate
Decay	n or τ_n	Moderate
Constant injection rate	$Q^{\dagger}$	Moderate
Constant concentration	Q	Relatively high

Note: † The volume has no influence only when the airflow rate, Q, is constant.

Table 1.2 *Summary of single-zone methods*

Method name	Tracer injection technique	Interpretation method	Suited for	
			Unbiased average	Continuous record
Transient methods (tracer gas concentration changes)				
(Simple) decay	Decay	Identification	No	No
Two-point decay	Decay	Integral	Yes	(Yes)
Step-up	Constant rate	Identification	No	No
Steady-state methods (tracer gas concentration is nearly constant)				
Pulse	Pulse	Integral	No	(Yes)
Constant injection	Constant rate	Direct solution	Yes[†]	No
Long-term integral	Any	Integral	No	No
Constant concentration	Constant concentration	Integral	Yes	Yes

Note: † Under condition.
Source: Sherman, 1990.

- The long-term integral method, generally used with passive sources and samplers, also gives a biased estimate of the average airflow rate. Since the measurement time is larger, the bias may not be negligible. This technique, however, provides an unbiased estimate of the average tracer concentration. If the tracer is used to simulate a contaminant, such experiments are of great interest for indoor air quality studies.
- The constant concentration technique is accurate and gives an unbiased estimate of the average airflow rate, but it requires the most technical equipment.

Table 1.3 *Summary of multi-zone methods*

Tracer injection strategy	Unbiased average of time-varying flow	Well suited for continuous record
Single tracer *(repeated measurements)*		
Pulse injection	No	No
Decay	No	No
Constant injection rate	No	No
Constant concentration	Yes	No
Multi-tracer		
Pulse injection	No	No
Decay	Yes*	No
Constant injection rate	No	Yes
Constant concentration	Yes	Yes

Note: *Only for a two-point estimate.

- Constant injection used with long-term direct solution is simpler to use and may give, under certain conditions, unbiased estimates of an average airflow rate.

The two-point decay method, and more generally the multi-zone, transient methods may lead to unacceptably large uncertainties if the measurement time period is inappropriate. See Enai *et al.* (1990) for two-zone, two-tracer, step-up and decay methods.

2

Airflow Rates
in Air Handling Units

Air handling units are designed to supply new air to the ventilated zone and to extract vitiated air from this zone. Many other airflows may be found in such units, as shown in Figure 2.1.

Measurements of airflow rates in ventilation systems are useful in order to check if the air follows the expected paths and thus detect potential problems early so they can be corrected, also allowing the optimization of the performance of the airflow system. Checks include examining if actual airflow rates are close to the desired values and if leakages and short-circuits are negligible. The checks should be performed:

- when commissioning a new ventilation system in order to ensure that the system is built according to the design specifications;
- if there are indoor air quality problems to help finding their causes;
- before refurbishing a ventilation system in order to accurately identify the potential problems to be cured by the refurbishment.

Commissioning mechanical ventilation systems is paramount in order to ensure that they function as designed. This includes not only the measurement of the main airflow rates and pressure distributions, but also checks that there are no excessive leakages or shortcuts. It should be noted that commissioning protocols are available for most units mentioned in the Introduction that function as designed, while they are not available for units in which airflow rates are not those designed or for units showing significant leakages or shortcuts.

This chapter proposes methods for measuring most airflow rates that occur in ventilation systems.

Measurement of the airflow rate in a duct

Summary of measurement methods

Common methods used to measure airflow rates in ducts exploit well-known relationships between flow rate and pressure drop across a restricted section

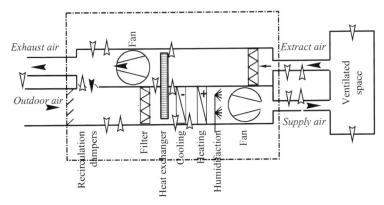

Figure 2.1 *Schematics of a supply and exhaust air handling unit*
Note: The main airflow paths are shown as solid arrows, and secondary or parasitic airflow paths are shown as open arrows.
Source: Roulet *et al.*, 2000a.

placed in the flow, for example a nozzle, Venturi or sharp-edged orifice (ISO, 2003). Alternatively, the air speed can be measured directly at a number of points lying in a cross-section of the duct (a traverse), and the results integrated along the traverse to give the volume flow rate (ISO, 1977). All of these methods have the disadvantage that a long straight section of duct, both upstream and downstream of the measurement point, is needed in order to condition the flow. Moreover, the introduction of a restriction may significantly change the airflow rate to be measured. Tracer techniques (ISO, 1978; Axley and Persily, 1988; Riffat and Lee, 1990), which avoid these problems, employ gas analysers and measure the dilution of a tracer gas introduced into the flow, using equipment that is becoming increasingly common, robust and easy to use.

Orifice plate, nozzle and Venturi flowmeters

The change in pressure in a pipe with a section of restricted area can be calculated by the Bernoulli law, provided there is neither friction nor compression. A relationship can hence be found between a pressure difference along the flow line and the corresponding flow rate, which may itself be deduced from a differential pressure measurement.

However, since there is a slight friction, the mass flow is:

$$Q = C_d A \sqrt{\frac{2\Delta p}{1 - \beta^2}} \tag{2.1}$$

where:

C_d is the discharge coefficient, taking account of friction losses,
A is the smallest cross-section in the flow,
Δp is the pressure difference between two taps properly located,

β is the reduction ratio, which is the ratio of the smallest diameter to the diameter of the pipe.

The flow may be restricted with an orifice plate, a nozzle or a Venturi tube. The most sophisticated and expensive is the Venturi tube in which the discharge coefficient is nearly 1 and constant for $Re > 2 \times 10^5$ and higher than 0.94 if $Re > 50{,}000$. Moreover, this device does not induce a large pressure drop in the flow. At the other end of the spectrum is the simple and cheap orifice plate, which induces a large pressure drop and shows discharge coefficients that may be as low as 0.6, depending on the Reynolds number. The characteristics of the nozzle lie in between. The device that is most sensitive to perturbation is the orifice, then the nozzle. The Venturi tube is the least sensitive.

All these flowmeters should be mounted between two straight pipes, the upstream pipe being up to 30 pipe diameters long, depending on the type of perturbation upstream, and the downstream part at least 3 diameters long. If a straightening vane 2 diameters long is installed upstream, the distance between this vane and the flowmeter may be reduced down to 10 diameters. The literature (for example, ASHRAE, 2001) provides detailed drawings of such devices.

Velocity traverse

If the velocity of the air, v, is measured at enough points in the duct, the volume airflow rate can be deduced by integration over the whole area, A, of the cross-section as shown:

$$Q = \rho \int_A v \, dA \tag{2.2}$$

where ρ is the density of the air.

Equipment to measure air speed

Provided the direction of flow is parallel to the duct, which may normally be assumed, then velocity can be determined by the measurement of air speed alone.

The air speed measurement devices should be small enough to enable them to be easily inserted through small holes in duct walls. The most common examples are hot wire or NTC anemometers, helix anemometers and Pitot tubes.

The hot wire anemometer and NTC anemometers measure the temperature drop of a heated wire or a heated resistor (with a negative temperature coefficient), which, in each case, is directly related to the temperature and speed of the air flowing over it. The sensors are heated by an electric current and measurements are made of the voltage drop, which depends on the temperature. The temperature of still air is taken into account by the use of a reference sensor shielded from the flow. Such devices can measure speeds

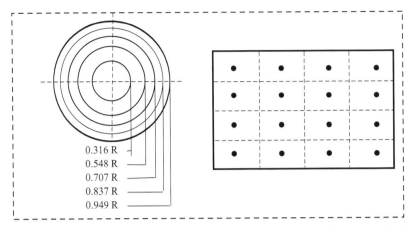

0.316 R
0.548 R
0.707 R
0.837 R
0.949 R

Figure 2.2 *Location of the measurement points in circular and rectangular ducts*
Source: ASHRAE, 2001.

from 0.05 to 5 m/s, and are well suited for speeds of 1–5 m/s, which are typical in ventilation ducts.

Helix anemometers measure the rotation speed of a small helix that is placed perpendicular to the airflow. The pressure difference between the front and the side of the Pitot tube is proportional to the square of the air speed. The helix anemometer and the Pitot tube are most accurate for air speeds above 10 m/s, and therefore may not be the best for measurement in ducts, where speeds as low as 1 m/s can be measured.

Test procedure

The location of the measurement in the duct should be at least 8 diameters downstream and 3 diameters upstream of any disturbances in the flow, such as a bend in the duct or a change in cross-section. Flow-straightening vanes located at 1.5 diameters upstream will improve the measurement accuracy.

Several measurements across the duct should be taken to enable integration. It is advisable to notionally divide the duct section into sub-sections of equal area and to take measurements at their centres. The ASHRAE *Handbook of Fundamentals* (ASHRAE, 2001) proposes a division of 16 to 64 rectangular sub-sections for a rectangular duct, and 20 annuli for cylindrical ducts (see Figure 2.2). In the latter, any asymmetry in the flow may be taken into account by taking measurements along two orthogonal directions.

Tracer gas dilution

Measurement of airflow rate in a duct is the simplest application of the tracer gas dilution technique. It is illustrated in Figure 2.3. The tracer is injected at a known constant flow rate, I. The air is analysed downstream, far enough from the injection port to have a good mixing of the tracer into the air (see 'Sampling points for concentration measurements', below).

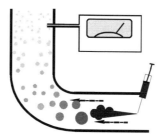

Figure 2.3 *Measuring the airflow rate in a duct with the tracer gas dilution method*

Source: Awbi, 2007.

Assuming that no tracer is lost in between, the mass balance of the tracer gas is, at steady state:

$$I = (C - C_0)Q \tag{2.3}$$

where:

C is the tracer concentration as obtained by analysis,
C_0 the concentration upstream the injection port (if any),
Q the airflow rate in the duct, which is then:

$$Q = \frac{I}{(C - C_0)} \tag{2.4}$$

This simple method assumes steady state: both airflow rate and injection flow rate are constant and the concentration is recorded only when a constant concentration is reached. This time is rather short, about five times the time period needed for the first molecules of tracer gas to reach the analyser if the air is not recirculated after delivery into a room. If there is recirculation, the time needed to reach the steady state may be much longer, i.e. five times the nominal time constant of the ventilated space. In this case, applying the method described in 'Measurement of airflow rate in a duct' (above) will reduce the time needed for measurements.

Note that, in a first approximation, mass balance can be replaced by volume balance to get volume flow rates instead of mass flow rates. This approximation remains valid as long as the difference between indoor and outdoor temperature is smaller than 1°C.

Airflow measurements at ventilation grilles

Ventilation inlet grilles should distribute the air in each room or space. In supply and exhaust units, extract grilles should absorb an equivalent airflow rate. Airflows should be balanced so that each room or space receives an appropriate airflow rate. Measurements at each grille are essential to ensure such a balance.

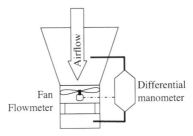

Figure 2.4 *Schematics of a compensated flowmeter*
Note: The differential manometer adjusts the fan so that there is no pressure drop through the flowmeter.

The methods described in the section on 'Measurement of airflow rate in a duct' (above) may of course be used to measure the airflow rates through grilles, but specific instruments may be easier to use or bring a more accurate result.

Inflatable bag

This method consists in measuring (using a chronograph) the time required to fill a plastic bag of a known volume when its opening is placed against the grid. It is not very accurate and creates a counter-pressure that could perturb the airflow, but the equipment is very cheap: a simple plastic bag, such as a rubbish bag.

Flowmeter

Any type of flowmeter fixed to a box or cone adapted to the shape of the grille may be used to measure the airflow rate through the grille, provided this flowmeter does not require too large a pressure drop across it.

Compensated flowmeter

The compensated flowmeter (see Figure 2.4) is equipped with a fan, the speed of which is adjusted so that the pressure differential through the instrument is negligible during the measurement. This allows an accurate measure of the flow rate going through the instrument without perturbing the measured grid or duct. The fan itself could be the flowmeter, since the fan speed at zero pressure differential is directly related to the airflow rate.

Airflow rate measurements in air handling units

Tracer gas injection ports

Injecting several tracer gases at various locations and analysing the air at other appropriate locations allows the assessment of several airflow rates simultaneously.

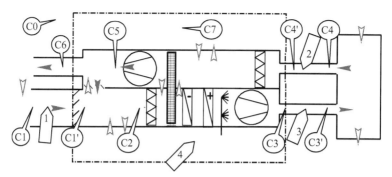

Figure 2.5 *Locations of tracer gas injection (arrows), and sampling points for concentration measurements (C_i) in a typical supply and exhaust air handling unit. Source*: Roulet *et al.*, 1999.

In principle, the method described above in 'Tracer gas dilution' can be applied to each branch of a duct network. However, this requires as many tracer gas injections and air sampling measurements as there are airflow rates. Experience has shown that the experiment can be made simpler, as shown in Figure 2.5, where most practical and efficient injection locations are indicated by arrows. In this figure, air sampling points for the required tracer gas concentration measurements are also shown.

If several tracer gases are needed but not available, it is possible to use the same tracer gas in several experiments, injecting the tracer successively at different locations. In this case, care should be taken to ensure constant airflow rates in the system. In particular, frequency controllers of the fans should be blocked at a constant frequency. It is also recommended to start with injection at location 2, then 3 and finally 1. This strategy shortens the time required between two experiments to reduce the tracer gas concentration in the system to a negligible level.

Two tracer gases or two successive measurements with one tracer gas allow in most cases assessment of all primary and most secondary airflow rates:

- tracer one injected in the main return air duct;
- tracer two (or a second run with tracer one) injected in the outside air duct.

Additional injection ports may be useful to increase the accuracy. These are:

- tracer three (or a third run with tracer one) injected in the main supply air duct, allowing the direct and more accurate determination of supply airflow rate;
- tracer four (or a fourth run with tracer one) injected in the control room at constant concentration to determine leakage from the control room into the air handling unit.

The optimal tracer gas injection rate depends on the design airflow rate Q_o in the duct and on the required concentration, C, itself depending on the sensitivity of the tracer gas analyser. A good method is to adjust the tracer gas injection flow rate on the basis of the outdoor airflow rate Q_{01}. If C_k is the expected tracer gas concentration of tracer k:

$$I_k = C_k Q_{01} \tag{2.5}$$

Sampling points for concentration measurements

Tracer gas concentrations are measured at several carefully chosen locations in order to obtain enough information to determine all the wanted airflow rates. It is important that there is a good mixing of tracer gas in the measured airflow. For this, several criteria should be fulfilled. Practice has shown that sufficient mixing is reached when the distance between injection ports and air sampling location is at least:

- 10 diameters (or duct widths) in straight ducts;
- 5 diameters if there is a mixing element such as bends, droplet catcher or a fan between injection ports and the air sampling location.

Proper mixing can be checked by looking at the variations of measured concentration with time, and when displacing the sampling location within the duct (see Figure 2.6). If variations are large and random, change the sampling and/or injection points, or use multiple injection ports until variations are within the usual measurement noise.

If the minimum distances mentioned above cannot be achieved, use multiple injection (see Figure 2.7) or install obstacles in the airflow to increase the turbulence.

Turbulent flow may transport some tracer gas a little upwind of the injection point. Therefore, the distance between sampling location upwind of injection points and the injection nozzle should be at least one duct diameter when there is no possibility of backward flow, and larger (3–5 duct diameters) when backward airflow is suspected (for example close to T junctions).

When sampling, never use the tubes that were once used for injecting a pure tracer gas, since some gas absorbed in the plastic of the tubes may be desorbed,

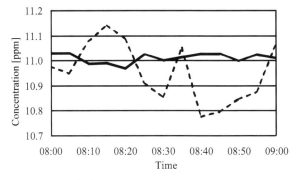

Figure 2.6 *Evolution of tracer gas concentration versus time*
Note: The solid line indicates good mixing of the tracer gas; the broken line indicates poor mixing.

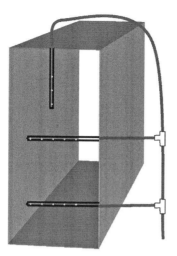

Figure 2.7 *Example of multiple injection devices*

bringing additional tracer into the analyser, thus biasing the concentration measurement. To avoid this, use different colours for injection and sampling tubes.

Principle of the interpretation procedure

The ductwork is modelled by nodes connected by ducts. In principle, the same equations as those used in Chapter 1, 'Application to buildings, multi-zone', to assess airflow rates measurements in multi-zone buildings could be used, the nodes being considered as zones. There is, however, an important difference since, in many cases, the directions of inter-zone airflows are known, leading to important simplifications in the system of equations.

In addition, tracer gas and air mass conservation equations can be written for each node in the duct network, and this provides, in most cases, a number of equations much larger than the number of unknown airflow rates (see for example 'Building the system of equations', below). There are several ways to use this peculiarity, which are described below for information. We have nevertheless found that in practice the most robust system of equations (the system that is the least sensitive to measurement uncertainties) can easily be purpose-built for each type of air handling unit, as shown in 'Simplest way', below.

Node by node method

The method presented in Chapter 1, 'Zone by zone systems of equations', can also be applied to ductwork and air handling units. Airflow and tracer gas conservation equations can be rearranged so as to obtain one system of equations per node, giving all airflow rates entering in this node. At steady state

$$-I_{ik} = \sum_{j=0}^{N} [C_{jk} - C_{ik}] Q_{ji} \qquad (2.6)$$

where:

I_{ik} is the injection rate of tracer gas k in (or just upwind of) node i,
C_{jk} is the concentration of tracer gas k in (or just downwind of) node j,
Q_{ji} is the airflow rate from node j to node i.

'Just upwind' and 'just downwind' mean far enough from the node to ensure a good mixing, but close enough to have no branching between the injection port or sampling location and the node.

Each system can be rewritten in a matrix form:

$$\vec{I}_i = \mathbf{C}_i \vec{Q}_i \qquad (2.7)$$

where:

$\vec{I}_i$ is the vector containing the tracer gas injection rate in the zone i,
$\mathbf{C}_i$ is the matrix containing the concentrations differences, $C_{jk} - C_{ik}$, of tracer k between zones j and i,
$\vec{Q}_i$ is the vector of airflow rates entering into zone i from zones j.

Airflow rates leaving the zones are determined by mass conservation equations

$$Q_{i0} = \sum_{j=0}^{N} [1 - \delta_{ij}] Q_{ji} - \sum_{j=1}^{N} [1 - \delta_{ij}] Q_{ij} \qquad (2.8)$$

An application to a typical air handling unit is presented below.

General method for 'black box' air handling unit

In most cases, it is not practical to inject tracer gases and to sample the air within the air handling unit. It is often much easier to find (or to bore) small holes in duct walls to insert the injection and sampling tubes. Therefore, a method for assessing airflow rates in air handling units using injection and sampling ports located only outside the units is presented below.

Building the system of equations

The ducts, leakage and shortcut network in the air handling system seen from outside, like a black box, are represented schematically in Figure 2.8.

Recirculation may be on purpose, or could result from leakage such as that sometimes found in heat exchangers. It occurs anyway between nodes 6 and 2. Regarding indoor air quality, there is no difference whether the

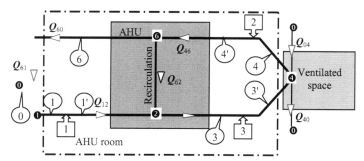

Figure 2.8 *The simplified network representing the air handling unit and ducts*
Note: Numbers in black circles represent the nodes of the network; boxes with arrows
are tracer gas injection locations; and numbered balloons are air sampling locations.
Arrows represent possible airflow rates.
Source: Awbi, 2007.

recirculated air passes through a leak between extract and supply parts of the
air handling unit or through a purpose-installed duct. Alternatively, ventilation units with heat exchangers seldom have recirculation ducts. Therefore,
the simplified network, as shown in Figure 2.8, is adapted for most
investigations.

The possible airflows are shown in Table 2.1.

Using four tracer gases as illustrated in Figure 2.8 and writing the conservation equations for them at the nodes gives the following system of equations:

Node 1, air inlet

$$0 = (C_{01} - C_{11})Q_{01} + (C_{61} - C_{11})Q_{61} \qquad (2.9)$$

$$0 = (C_{02} - C_{12})Q_{01} + (C_{62} - C_{12})Q_{61}$$

$$0 = (C_{03} - C_{13})Q_{01} + (C_{63} - C_{13})Q_{61}$$

$$0 = (C_{04} - C_{14})Q_{01} + (C_{64} - C_{14})Q_{61}$$

Table 2.1 *Possible airflow rates in the network represented in Figure 2.8*

		Going into node					
		0	1	2	4	6	7
Coming from node	0		$\mathbf{Q_{01}}$		Q_{04}		Q_{07}
	1			$\mathbf{Q_{12}}$			
	2				$\mathbf{Q_{24}}$	Q_{26}	Q_{27}
	4	Q_{40}				$\mathbf{Q_{46}}$	
	6	$\mathbf{Q_{60}}$	Q_{61}	$\mathbf{Q_{62}}$			Q_{67}
	7	Q_{70}		Q_{72}		Q_{76}	

Note: The main airflows are in bold. The others are parasitic airflow rates that in principle should
be negligible.

Node 2, return

$$-I_{11} = (C_{11} - C_{31})Q_{12} + (C_{61} - C_{31})Q_{62} + (C_{71} - C_{31})Q_{72} \qquad (2.10)$$
$$0 = (C_{12} - C_{32})Q_{12} + (C_{62} - C_{32})Q_{62} + (C_{72} - C_{32})Q_{72}$$
$$0 = (C_{13} - C_{33})Q_{12} + (C_{63} - C_{33})Q_{62} + (C_{73} - C_{33})Q_{72}$$
$$0 = (C_{14} - C_{34})Q_{12} + (C_{64} - C_{34})Q_{62} + (C_{74} - C_{34})Q_{72}$$

Node 4, vented space

$$0 = (C_{01} - C_{41})Q_{04} + (C_{31} - C_{41})Q_{24} \qquad (2.11)$$
$$0 = (C_{02} - C_{42})Q_{04} + (C_{32} - C_{42})Q_{24}$$
$$-I_{43} = (C_{03} - C_{43})Q_{04} + (C_{32} - C_{42})Q_{24}$$
$$0 = (C_{04} - C_{44})Q_{04} + (C_{34} - C_{44})Q_{24}$$

Node 6, recirculation

$$0 = (C_{31} - C_{61})Q_{26} + (C_{41} - C_{61})Q_{46} + (C_{71} - C_{61})Q_{76} \qquad (2.12)$$
$$-I_{62} = (C_{32} - C_{62})Q_{26} + (C_{42} - C_{62})Q_{46} + (C_{72} - C_{62})Q_{76}$$
$$0 = (C_{33} - C_{63})Q_{26} + (C_{43} - C_{63})Q_{46} + (C_{73} - C_{63})Q_{76}$$
$$0 = (C_{34} - C_{64})Q_{26} + (C_{44} - C_{64})Q_{46} + (C_{74} - C_{64})Q_{76}$$

Node 7, technical room

$$0 = (C_{01} - C_{71})Q_{07} + (C_{31} - C_{71})Q_{27} + (C_{61} - C_{71})Q_{67} \qquad (2.13)$$
$$0 = (C_{02} - C_{72})Q_{07} + (C_{32} - C_{72})Q_{27} + (C_{62} - C_{72})Q_{67}$$
$$0 = (C_{03} - C_{73})Q_{07} + (C_{33} - C_{73})Q_{27} + (C_{63} - C_{73})Q_{67}$$
$$-I_{74} = (C_{04} - C_{74})Q_{07} + (C_{34} - C_{74})Q_{27} + (C_{64} - C_{74})Q_{67}$$

At each node, the entering mass of air equals the leaving mass. Taking account of the possible airflows given in Table 2.1, we get, after some reorganization and leaving main airflow rate at the left-hand side and parasitic airflow rates at the right-hand side of each equation:

Node 0, outdoors $\quad Q_{01} - Q_{60} = Q_{40} - Q_{04} + Q_{70} - Q_{07}$ \qquad (2.14)

Node 1, inlet $\quad Q_{01} - Q_{12} = -Q_{61}$ \qquad (2.15)

Node 2, return $\quad Q_{12} - Q_{24} + Q_{62} = Q_{26} + Q_{27} - Q_{72}$ \qquad (2.16)

Node 4, vented space $\quad Q_{24} - Q_{46} = Q_{40} - Q_{04}$ \qquad (2.17)

Node 6, recirculation $\quad Q_{46} - Q_{60} - Q_{62} = -Q_{26} + Q_{61} + Q_{67} - Q_{76}$ (2.18)

Node 7, technical room $\quad Q_{07} - Q_{70} = Q_{72} - Q_{27} + Q_{76} - Q_{67}$ \qquad (2.19)

This system of 27 equations when combined with the system of Equation 2.9 can be solved in various ways to provide the six main airflow rates and potentially ten parasitic flow rates. This global system of equations contains more equations than unknowns. There are several ways to address this situation, and we have found that some methods are better than others for application to air handling units. Therefore we present the tested methods below.

Least square solution

The system of equations from 2.9 to 2.19 is over-determined: there are 26 equations for calculating 16 airflow rates. In zones where $\vec{I}_i \neq 0$, the system could be solved by least square fit:

$$\vec{Q}_i = [\boldsymbol{C}^T \quad \boldsymbol{C}_i^T]^{-1} \boldsymbol{C}_i^T \vec{I}_i \qquad (2.20)$$

where $\boldsymbol{C}^T$ is $\boldsymbol{C}$ transposed. The resulting flow vector is the one that best satisfies the set of equations. However, the injection rate vector $\vec{I}'$, back-calculated using Equation 2.7 with the resulting flow vector $\vec{Q}$ and the measured concentration will not be equal to the actual one. This method always provides a solution, but, depending on the condition of the system of equations, this solution could be far from the reality.

At nodes where the tracer i is not injected, the system can only provide linear combinations of airflow rates, as far as the determinant $|\boldsymbol{C}_i| = 0$.

Eliminating some equations

Combining some of the equations of system 2.9 to 2.19 allows avoidance of the measurement of some concentrations. A system having as many equations as unknown airflow rates can be solved using:

$$\vec{Q} = \boldsymbol{C}^{-1} \vec{I} \qquad (2.21)$$

Experience showed that this interpretation method often leads to poorly conditioned systems of equations. Results are then very sensitive to slight changes of input data.

Looking for the best conditioned system

A set of N equations (N being the number of unknown airflow rates, in this case 16) can be selected to give the best accuracy, or the smallest sensitivity to variations or errors of injection rates and concentrations. This set can be theoretically selected by calculating the condition number (see Chapter 3, 'Condition of the model matrix') of all possible sets of equations extracted from the full system, and taking the set with the smallest condition number. This selection could be tedious: there are 13,037,895 sets of 16 equations that can be extracted from the system 2.9!

Simplest way

A method providing all airflow rates with the simplest solutions – hence probably the least sensitive to measurement errors – is given below.

The results are provided with their confidence intervals, calculated under the assumption that random and independent errors affect the measurements of tracer gas concentration and injection rates. In this case, the confidence interval of any result, for example an airflow rate, is (see Chapter 7, 'Error analysis'):

$$[Q - \delta Q; Q + \delta Q] \quad \text{with} \quad \delta Q = T(P, \infty) \sqrt{\sum_i \left(\frac{\partial Q}{\partial x_i}\right)^2 \delta x_i^2} \tag{2.22}$$

where:

$T(P, \infty)$ is the Student coefficient for having the actual value within the confidence interval with probability P,

x_i is for all variables (other airflow rates, tracer gas concentration and injection rates) on which Q depends,

δx_i is for the standard deviation of the variable x_i, assumed to be a random variable of mean x_i.

The various airflow rates are then:

Intake airflow rate

$$Q_{12} = \frac{I_{11}}{C_{1'1} - C_{11}} \tag{2.23}$$

with

$$\delta Q_{12} = T(P, \infty) \sqrt{\frac{(C_{1'1} - C_{11})^2 \delta I_{11}^2 + I_{11}^2 (\delta C_{1'1}^2 + \delta C_{11}^2)}{(C_{1'1} - C_{11})^4}} \tag{2.24}$$

Supply airflow rate

$$Q_{24} = \frac{I_{43}}{C_{3'3} - C_{33}} \tag{2.25}$$

with

$$\delta Q_{24} = T(P, \infty) \sqrt{\frac{(C_{3'3} - C_{33})^2 \delta I_{43}^2 + I_{43}^2 (\delta C_{3'3}^2 + \delta C_{33}^2)}{(C_{3'3} - C_{33})^4}} \tag{2.26}$$

Note that $C_{6k} = C_{4'k}$, as long as there are no leakages into the air handling unit between locations $4'$ and 6.

Extract airflow rate

$$Q_{46} = \frac{I_{62}}{C_{4'2} - C_{42}} \tag{2.27}$$

with

$$\delta Q_{46} = T(P, \infty) \sqrt{\frac{(C_{4'2} - C_{42})^2 \delta I_{52}^2 + I_{52}^2 (\delta C_{4'2}^2 + \delta C_{42}^2)}{(C_{4'2} - C_{42})^4}} \tag{2.28}$$

Outdoor airflow rate

$$Q_{01} = Q_{12} \frac{C_{6k} - C_{1k}}{C_{6k} - C_{0k}} \tag{2.29}$$

where $k = 1$ or 2 is recommended.

$$\delta Q_{01} = \frac{T(P, \infty)}{(C_{6k} - C_{0k})^2} \sqrt{f_{01}} \tag{2.30}$$

where

$$f_{01} = (C_{6k} - C_{1k})^2 (C_{6k} - C_{0k})^2 \delta Q_{12}^2$$
$$+ Q_{12}^2 [(C_{6k} - C_{0k})^2 \delta C_{1k}^2 + (C_{6k} - C_{1k})^2 \delta C_{0k}^2 + (C_{1k} - C_{0k})^2 \delta C_{6k}^2]$$

External shortcut

$$Q_{61} = Q_{12} - Q_{01} = Q_{12} \left[1 - \frac{C_{61} - C_{11}}{C_{61} - C_{01}} \right] = Q_{12} \frac{C_{11} - C_{01}}{C_{61} - C_{01}} \tag{2.31}$$

$$\delta Q_{61} = \sqrt{\delta Q_{12}^2 + \delta Q_{01}^2} = \frac{T(P, \infty)}{(C_{61} - C_{01})^2} \sqrt{f_{61}} \tag{2.32}$$

where

$$f_{61} = (C_{61} - C_{01})^2 (C_{01} - C_{11})^2 \delta Q_{12}^2$$
$$+ Q_{12}^2 [(C_{61} - C_{01})^2 \delta C_{11}^2 + (C_{11} - C_{61})^2 \delta C_{01}^2 + (C_{01} - C_{11})^2 \delta C_{61}^2]$$

Recirculation flow rate

$$Q_{62} + \left[Q_{72} \frac{C_{7k} - C_{3k}}{C_{6k} - C_{3k}} \right] = Q_{12} \frac{C_{3k} - C_{1k}}{C_{6k} - C_{3k}} \tag{2.33}$$

Here again, it is recommended to use concentrations of tracer gas 2 ($k = 2$). Note that Q_{62} is aliased with Q_{72}, multiplied by a coefficient that is very small for tracer 2, since only C_{62} differs significantly from zero.

$$\delta Q_{62} = \frac{T(P, \infty)}{(C_{6k} - C_{3k})^2} \sqrt{f_{62}} \tag{2.34}$$

where

$$f_{62} = (C_{6k} - C_{3k})^2 (C_{3k} - C_{1k})^2 \delta Q_{12}^2$$
$$+ Q_{12}^2 [(C_{6k} - C_{3k})^2 \delta C_{1k}^2 + (C_{6k} - C_{1k})^2 \delta C_{3k}^2 + (C_{1k} - C_{6k})^2 \delta C_{6k}^2]$$

The bias resulting from the alias with Q_{72} is not taken into account in the confidence interval. If tracer 4 is used, we get:

Leakage to node 2

$$Q_{72} = \frac{(C_{32} - C_{12})(C_{64} - C_{34}) - (C_{34} - C_{14})(C_{62} - C_{32})}{(C_{64} - C_{34})(C_{72} - C_{32}) - (C_{74} - C_{34})(C_{62} - C_{32})} \tag{2.35}$$

$$\delta Q_{72} = \frac{T(P, \infty)}{(C_{64} - C_{34})(C_{72} - C_{32}) - (C_{74} - C_{34})(C_{62} - C_{32})} \sqrt{f_{72}} \tag{2.36}$$

where

$$\begin{aligned}
f_{72} = & (C_{64} - C_{34})^2 \delta C_{12}^2 + [C_{64} - C_{14} + Q_{72}(C_{64} - C_{74})]^2 \delta C_{32}^2 \\
& + [C_{34} - C_{14} + Q_{72}(C_{74} - C_{34})]^2 \delta C_{62}^2 + Q_{72}^2(C_{62} - C_{34})\delta C_{72}^2 \\
& + (C_{62} - C_{32})^2 \delta C_{14}^2 + [C_{12} - C_{62} + Q_{72}^2(C_{72} - C_{62})]^2 \delta C_{34}^2 \\
& + [C_{32} - C_{12} - Q_{72}(C_{72} - C_{32})]^2 \delta C_{64}^2 + Q_{72}^2(C_{62} - C_{32})\delta C_{74}^2
\end{aligned}$$

The recirculation flow rate can then be determined separately:

$$Q_{62} = \frac{(C_{32} - C_{12})(C_{74} - C_{34}) - (C_{34} - C_{14})(C_{72} - C_{32})}{(C_{74} - C_{34})(C_{62} - C_{32}) - (C_{72} - C_{32})(C_{64} - C_{34})} \tag{2.37}$$

$$\delta Q_{62} = \frac{T(P, \infty)}{(C_{74} - C_{34})(C_{62} - C_{32}) - (C_{72} - C_{32})(C_{64} - C_{34})} \sqrt{f_{62}} \tag{2.38}$$

where

$$\begin{aligned}
f_{62} = & (C_{74} - C_{34})^2 \delta C_{12}^2 + [C_{74} - C_{14} + Q_{62}(C_{74} - C_{64})]^2 \delta C_{32}^2 \\
& + [Q_{62}(C_{74} - C_{34})]^2 \delta C_{62}^2 + [C_{14} - C_{34} + Q_{62}(C_{64} - C_{34})]^2 \delta C_{72}^2 \\
& + (C_{62} - C_{32})^2 \delta C_{14}^2 + [C_{12} - C_{72} + Q_{62}^2(C_{62} - C_{72})]^2 \delta C_{34}^2 \\
& + [C_{32} + C_{12} - Q_{62}(C_{62} - C_{32})]^2 \delta C_{74}^2 + Q_{62}^2(C_{72} - C_{32})\delta C_{64}^2
\end{aligned}$$

Infiltration flow rate

$$Q_{04} = \frac{(C_{33} - C_{43})Q_{24} + I_{43}}{C_{43} - C_{03}} \tag{2.39}$$

with

$$\delta Q_{04} = \frac{T(P, \infty)}{(C_{43} - C_{03})^2} \sqrt{f_{04}} \tag{2.40}$$

where

$$\begin{aligned}
f_{04} = & (C_{43} - C_{03})^2 (C_{33} - C_{43})^2 \delta Q_{24}^2 + Q_{24}^2(C_{43} - C_{03})^2 \delta C_{33}^2 \\
& + [Q_{24}(C_{03} - C_{33}) + I_{43}]^2 \delta C_{43}^2 + [Q_{24}(C_{33} - C_{33}) + I_{43}]^2 \delta C_{03}^2
\end{aligned}$$

or

$$Q_{04} \cong Q_{24} \frac{(C_{3k} - C_{4k})}{(C_{4k} - C_{0k})} = Q_{12} \frac{(C_{6k} - C_{1k})}{(C_{6k} - C_{3k})} \frac{(C_{3k} - C_{4k})}{(C_{4k} - C_{0k})} \tag{2.41}$$

$$\delta Q_{04} = \frac{T(P, \infty)}{(C_{4k} - C_{0k})^2} \sqrt{f_{04}} \tag{2.42}$$

where

$$f_{04} = (C_{6k} - C_{3k})^2 (C_{3k} - C_{1k})^2 \delta Q_{12}^2$$
$$+ Q_{12}^2 [(C_{6k} - C_{3k})^2 \delta C_{1k}^2 + (C_{6k} - C_{1k})^2 \delta C_{3k}^2 + (C_{1k} + C_{6k})^2 \delta C_{6k}^2]$$

with $k \neq 3$ (recommended value: $k = 1$).

Exfiltration flow rate

$$Q_{40} = Q_{04} + Q_{24} - Q_{46} \tag{2.43}$$

with

$$\delta Q_{40} = \sqrt{\delta Q_{04}^2 + \delta Q_{24}^2 + \delta Q_{46}^2} \tag{2.44}$$

Inverse recirculation airflow rate through the air handling unit is:

$$Q_{26} + [Q_{27} - Q_{72}] = Q_{62} - Q_{24} + Q_{12} \tag{2.45}$$

with

$$\delta Q_{26} = \sqrt{\delta Q_{62}^2 + \delta Q_{24}^2 + \delta Q_{12}^2} \tag{2.46}$$

Exhaust airflow rate

$$Q_{60} + [Q_{70} - Q_{07}] = +Q_{04} - Q_{40} + Q_{01} \tag{2.47}$$

with

$$\delta Q_{60} = \sqrt{\delta Q_{04}^2 + \delta Q_{40}^2 + \delta Q_{01}^2} \tag{2.48}$$

If there is no exhaust duct (supply only systems), Q_{60} is zero, and Equation 2.47 can be used to get either an estimate of the net leakage rate between the technical room and outdoor environment, $[Q_{70} - Q_{07}]$, or, if this is zero, to calculate Q_{40}, or the net infiltration rate.

If $Q_{60} = 0$ (supply only system):

$$Q_{04} - Q_{40} + [Q_{70} - Q_{07}] = Q_{01} \tag{2.49}$$

Leakage from technical room to node 6 can be assessed by:

$$Q_{76} = \frac{(C_{32} - C_{62})Q_{26} + (C_{42} - C_{62})Q_{46} + I_{62}}{C_{72} - C_{62}} \tag{2.50}$$

or by

$$Q_{76} = \frac{(C_{3k} - C_{6k})Q_{26} + (C_{4k} - C_{6k})Q_{46}}{C_{7k} - C_{6k}} \tag{2.51}$$

with $k \neq 2$ ($k = 4$ is not recommended here).

$$\delta Q_{76} = \frac{T(P, \infty)}{C_{7k} - C_{6k}} \sqrt{f_{76}} \qquad (2.52)$$

where

$$f_{76} = Q_{26}^2 \delta C_{3k}^2 + Q_{46}^2 \delta C_{4k}^2 + (Q_{26} + Q_{46} - Q_{76})^2 \delta C_{6k}^2$$
$$+ (C_{3k} - C_{6k})^2 \delta Q_{26}^2 + (C_{4k} - C_{6k})^2 \delta Q_{46}^2 + \delta(k, 2) \cdot \delta I_{6k}^2$$

where: $k = 1$, 2 or 3; and the delta function $\delta(k, 2) = 1$ if $k = 2$ and 0 if $k \neq 2$.

Leakage airflow rates to the technical room can be obtained from system and equation (2.19).

Less than four tracer gases

If the tracer gas is injected at only one or two locations, the corresponding equations should be deleted. In this case, some of the airflow rates cannot be determined.

If only one tracer gas is injected at location 1, merely Q_{01}, Q_{12} and Q_{62} can be measured. It is nevertheless interesting to assess three airflow rates in one shot and with one tracer gas only!

When a tracer is injected just at location 2, Q_{45} and Q_{62} only can be measured.

If tracer 4 is not used, leakage from the technical room into the unit cannot be measured. Large leakage can nevertheless be detected from unexpected dilution of the other tracer gases in the unit.

If tracer 3 is not used, Q_{24}, aliased with several leakage flow rates (its value includes a linear combination of parasitic airflow rates), can nevertheless be calculated from:

$$Q_{24} + \left[Q_{26} + Q_{27} - Q_{72} \left(1 + \frac{C_{7k} - C_{3k}}{C_{6k} - C_{3k}} \right) \frac{C_{7k} - C_{3k}}{C_{6k} - C_{3k}} \right]$$
$$= Q_{12} + Q_{62} = Q_{12} \frac{C_{6k} - C_{1k}}{C_{6k} - C_{3k}} \qquad (2.53)$$

with

$$\delta Q_{24} = \sqrt{\delta Q_{12}^2 + \delta Q_{62}^2} = \frac{T(P, \infty)}{(C_{6k} - C_{3k})^2} \sqrt{f_{24}} \qquad (2.54)$$

where

$$f_{24} = (C_{6k} - C_{1k})^2 (C_{6k} - C_{3k})^2 \delta Q_{12}^2$$
$$+ Q_{12}^2 [(C_{6k} - C_{3k})^2 \delta C_{1k}^2 + (C_{6k} - C_{1k})^2 \delta C_{3k}^2 + (C_{1k} - C_{3k})^2 \delta C_{6k}^2]$$

In this equation, $k = 2$ is recommended.

Planning tool

There are many types of air handling units, and, from our experience, each new measurement poses new problems. It is hence impossible to provide a detailed measurement protocol valid for all types. Therefore, a computer program was developed that performs the following tasks:

- Requests input data:
  - characterization of the air handling unit (type, location, design airflow rates, heat exchanger, position of fans with respect to heat exchanger, etc.);
  - tracer gas(es) used, injection location(s) and design concentration(s);
  - characterization of building (approximate volume, number of occupants, overpressurized or not, etc.);
  - airflows that obviously cannot exist.
- Evaluates the risk of poor tracer gas mixing from the distance between injection and sampling locations and from the devices (fans, bends, filters, dampers) placed in between.
- Prepares a printed measurement protocol containing injection and sampling locations, injection rates of tracer gases and a system of equation in accordance with 'Simplest way' (above).
- Requests measured tracer gas concentrations and actual injection flow rates or reads them in a file.
- Solves the system of equations and prepares a measurement report.

This piece of software is available on www.e4tech.com, in 'Software and Publications'.

Example of application

Sulphur hexafluoride was injected as tracer 1 and nitrous oxide as tracer 2 in an air handling unit without planned recirculation, but equipped with a rotating heat exchanger. Resulting concentrations are shown in Figure 2.10, and measurement results in Figure 2.9.

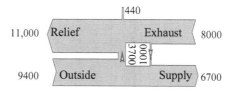

Figure 2.9 *Measured airflow rates in a leaky air handling unit*
Note: Design airflow rates were 13,000 m/h for both supply and return, and zero for recirculation.
Source: Awbi, 2007.

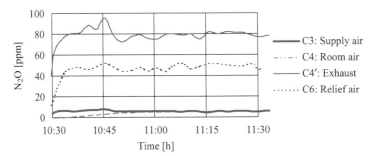

Figure 2.10 *Concentrations at locations shown in Figure 2.5 resulting from injection of SF_6 as tracer 1 and N_2O as tracer 2 in a leaky air handling unit*
Note: A shortcut through the heat exchanger dilutes exhaust air, thus decreasing the relief air concentration. The presence of this tracer gas in supply air results from parasitic recirculation.
Source: Awbi, 2007.

Leaks in the heat exchanger, as well as in the return air channel, were detected with this measurement. Measurement in three other identical units in the same office did not show any shortcut. However, measured outdoor airflow rates were between 55 and 66 per cent of the design value.

Simple measurement using CO_2 from occupants

A special case is when only one tracer is injected in the ventilated space. This could be the carbon dioxide emitted in the ventilated space by occupants. That tracer gas is of great practical interest since it does not need any injection system. In this case, Equation 2.33 can easily be solved. Assuming that there is no inverse recirculation, and no leaks in the air handling unit, the global recirculation rate is:

$$R = \frac{Q_{62}}{Q_{12} + Q_{62}} = \frac{C_{3k} - C_{1k}}{C_{4k} - C_{1k}} \tag{2.55}$$

with

$$\delta R = \frac{T(P, \infty)}{(C_{4k} - C_{3k})^2} \sqrt{f_R} \tag{2.56}$$

where

$$f_R = (C_{3k} - C_{4k})^2 \delta C_{1k}^2 + (C_{4k} - C_{1k})^2 \delta C_{3k}^2 + (C_{3k} - C_{1k})^2 \delta C_{4k}^2$$

And the equivalent outdoor airflow rate per occupant is:

$$\frac{Q_{01} + Q_{04}}{N_{persons}} = \frac{0.018 [m^3/h]}{C_{4k} - C_{0k}} \tag{2.57}$$

assuming that a person exhales 18 l/h of carbon dioxide and that occupants are the only indoor sources of CO_2. Airflow rates are in $m^3/(h\ person)$ if concentrations are in volumetric ratios. It is not possible with only one tracer injected into the ventilated space to differentiate between outdoor air from mechanical ventilation and from infiltration.

Measurements in buildings with large time constants

Most methods are designed to measure units with recirculation ratios below 50 per cent. This is the case of the method proposed above. However, air handling units designed to condition (heat or cool) spaces with large loads such as those found in cold or tropical climates often present large recirculation ratios that homogenize the concentrations, and large nominal time constants (ratio of the ventilated volume to the outdoor airflow rate) that strongly increase the time needed to get steady state in the supply duct (node 3) and in the room (node 4). There is, however, a way to shorten the measurement time by extrapolating the evolution of tracer gas concentration with time (Roulet and Zuraimi, 2003), which is described below.

Writing the conservation equation of tracer gas 3 at node 4, in the ventilated space, gives:

$$\rho V \frac{\partial C_{43}}{\partial t} = I_3 + Q_{24}C_{23} + Q_{04}C_{03} - (Q_{46} + Q_{40})C_{43} \tag{2.58}$$

Because of the large recirculation ratio, it can be assumed that the concentration is homogeneous in the ventilated space. Dividing this equation by the supply airflow rate Q_{24} gives:

$$\frac{\rho V}{Q_{24}} \frac{\partial C_{43}}{\partial t} = \frac{I_3}{Q_{24}} + C_{23} + \gamma_i C_{03} - (1 + \gamma_i)C_{43} \tag{2.59}$$

where γ_i is the infiltration ratio Q_{04}/Q_{24}. Using the definition of the nominal time constant τ_n, of the recirculation ratio R, and using the tracer gas conservation at node 2:

$$\frac{V}{Q_{24}} = \frac{V}{Q_{01}} \frac{Q_{01}}{Q_{24}} = \tau_n(1 - R) \quad \text{and} \quad C_{23} = R_{xs}C_{43} + (1 - R)C_{03} \tag{2.60}$$

we get

$$\tau_n(1 - R)\frac{\partial C_{43}}{\partial t} = \frac{\dot{m}_{t3}}{\dot{m}_{24}} - (1 - R + \gamma_i)(C_{43} - C_{03}) \tag{2.61}$$

The steady-state concentration is:

$$C_{43}^{\infty} = \frac{I_3}{Q_{24}(1 - R + \gamma_i)} + C_{03} \tag{2.62}$$

and

$$C_{43}(t) = C_{43}^{\infty}(1 - e^{t/\tau})$$

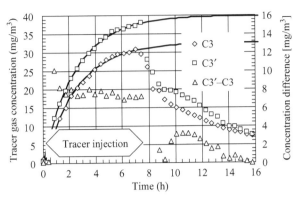

Figure 2.11 *Tracer gas concentrations in the supply duct, upstream (3) and downstream (3′) of the tracer gas injection port*

Note: Points are measured concentrations, while lines are exponential fits.
Source: Roulet and Zuraimi, 2003.

with

$$\tau = \frac{\tau_n(1 - R)}{1 - R + \gamma_i} \tag{2.63}$$

The theoretical exponential can be fitted to the experimental points, as shown in Figure 2.11 depicting an actual experiment. The concentration increase in the supply duct, $C_3' - C_3$, quickly reaches its steady-state value, while a good fit of the exponential can be obtained within two time constants, allowing the determination of the steady-state concentration and time constant without waiting for equilibrium that is reached after at least five time constants.

Appropriate method for assessing the recirculation ratio

In Equations 2.33, 2.34 and 2.41 the concentration difference $C_{6k} - C_{3k}$ is at the denominator, and these two concentrations are close to each other at steady state when the recirculation ratio is high. This leads to a large confidence interval of the calculated airflow rates. In units with large recirculation ratios, it is better to inject the tracer gas at location 3 instead of location 2. The supply airflow rate can then be determined with better accuracy, using:

$$Q_{24} = \frac{I_3}{C_{3'3} - C_{33}} \tag{2.64}$$

Assuming that the confidence interval is the same for both concentrations, the confidence interval is:

$$\frac{\delta Q}{Q} \cong T(P, \infty)\sqrt{\left(\frac{\delta I}{I}\right)^2 + 2\left(\frac{\delta C}{C' - C}\right)^2} \tag{2.65}$$

The recirculation airflow rate can then be calculated using:

$$Q_{62} = Q_{24} - Q_{12} \tag{2.66}$$

with

$$\delta Q_{62} = T(P, \infty)\sqrt{\delta Q_{24}^2 + \delta Q_{12}^2} \cong T(P, \infty)\sqrt{1 + (1 - R)^2}\, \delta Q \tag{2.67}$$

assuming that the relative error $\delta Q / Q$ is the same for both airflow rates, and taking into account that $Q_{12} = (1 - R)Q_{24}$. Note that, in this case, δQ_{62} decreases when R increases.

The extract airflow rate Q_{46} cannot be assessed without injecting a tracer gas in the extract duct. However, in air handling units that have no exhaust duct (such as most units in Singapore and other tropical countries), $Q_{60} = 0$, hence $Q_{46} = Q_{62}$, and $Q_{40} = Q_{01} + Q_{04}$.

The recirculation ratio is defined by:

$$R = \frac{Q_{62}}{Q_{24}} = \frac{Q_{62}}{Q_{62} + Q_{12}} \tag{2.68}$$

Assuming that there is no leak in the air handling unit, it can be assessed using three different methods:

Method A

$$R = \frac{C_{3k} - C_{1'k}}{C_{6k} - C_{1'k}} \tag{2.69}$$

the subscript k being for any tracer gas except the one injected in the inlet duct. The confidence interval is:

$$\delta R = \frac{T(P, \infty)}{(C_{6k} - C_{1'k})^2} \sqrt{f_R} \tag{2.70}$$

where

$$f_R = (C_{3k} - C_{6k})^2 \delta C_{1'k}^2 + (C_{6k} - C_{1'k})^2 \delta C_{3k}^2 + (C_{3k} - C_{1'k})^2 \delta C_{6k}^2 \tag{2.71}$$

If we assume that the relative error is the same for all concentrations, and taking into account that, for tracers injected at locations 2 and 3, $C_{1'k} \cong 0$ and therefore $C_{3k} \cong RC_{6k}$, we can get a simpler expression for the confidence interval of the recirculation ratio:

$$\delta R \cong \frac{T(P, \infty)\delta C}{C} \sqrt{2(R^2 - R + 1)} \tag{2.72}$$

The recirculation ratio can also be calculated using:

Method B

$$R = \frac{Q_{62}}{Q_{24}}$$

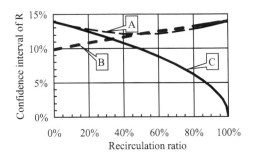

Figure 2.12 *Confidence interval of the recirculation ratio as a function of the recirculation ratio itself for three assessment methods*

Note: For this figure, the relative confidence intervals of injection rate and concentrations are 5 per cent.

with

$$\delta R = \frac{\delta Q}{Q}\sqrt{(1 + R)} \tag{2.73}$$

Or method C

$$R = 1 - \frac{Q_{12}}{Q_{24}}$$

with

$$\delta R = \sqrt{2}\frac{\delta Q}{Q}(1 - R) \tag{2.74}$$

assuming that the relative error $\delta Q/Q$ is the same for both airflow rates, and taking into account that $Q_{12} = (1 - R)Q_{24}$.

The accuracy of the three methods for determining R is compared in Figure 2.12. Method B (Equation 2.73) should be preferred at low recirculation ratios, while method C (Equation 2.74) is best at large recirculation ratios. Method A could be applied where the other methods cannot be used. Note that the relative error $\delta R/R$ becomes very large for small recirculation ratios.

It is interesting to see that some interpretation methods of the same measurements provide more accurate results than others, and that the best way depends on the unit measured. Therefore, care should be taken to select the most appropriate method.

3

Age of Air and Ventilation Efficiency

The airflow patterns should, in principle, be organized in order that new air is brought to the head of the occupants, so that they get fresh, clean air, and that contaminants be evacuated as quickly as possible, before being mixed with indoor air. However, air, as any other fluid, always follows the easiest path. This means that the airflow does not necessarily follow expected patterns. Since air is transparent, unexpected airflow patterns are noticed only when things go wrong. Depending on the airflow distribution in rooms and for a given airflow rate, the concentration of contaminants in the occupied space may vary by a factor of two or more. Therefore, measurements may be useful to check if the airflow patterns are as expected. Such measurements allow checks to ensure that:

- the air change efficiency is as large as possible,
- clean air is supplied to the right places,
- air contaminants are quickly removed.

The measurement of the age of the air allows for the detection of possible shortcuts and dead zones and for the checking of the general airflow pattern in the room or in a building.

Definitions

The quantities defined below are explained in greater detail elsewhere (Sandberg, 1984; Sutcliffe, 1990) and are only briefly described here.

Age of the air

Let us assume that the molecules of outdoor air start their indoor life when entering the building or the ventilation system. These arrive at a given location, r, in a room after a time, τ_r, that varies from one particle to the other. τ_r is called the residence time of the particle in the room, or its age. Note that the air elements themselves, i.e. oxygen and nitrogen molecules, do not age.

However, the more time a small volume of air spends in a room, the more it will be contaminated by pollutants.

Since there is a large number of air particles all taking different paths, we define a probability density $f(\tau_r)$ that the age of particles arriving at a given location is between τ and $\tau + d\tau$ and, a probability $F(\tau_r)$ that this age is larger than τ. These two functions are, by definition, related by:

$$f(\tau_r) = -\frac{dF(\tau_r)}{d\tau} \quad \text{and} \quad F(\tau_r) = 1 - \int_0^\tau f(t_r)\, dt \tag{3.1}$$

The local mean age of air at a point r is defined by the average age of all the air particles arriving at that point:

$$\bar{\tau}_r = \int_0^\infty t f_r(t)\, dt = \int_0^\infty F_r(t)\, dt \tag{3.2}$$

The room mean age of air $\langle \tau \rangle$ is defined by the average of the ages of all the air particles in the room.

Nominal time constant

The nominal time constant, τ_n, of a ventilated zone, is the ratio of its volume, V, to the supplied fresh volume airflow rate, q, (including infiltration), or the ratio of the mass of air contained in the space, M, to the mass airflow rate, Q:

$$\tau_n = \frac{V}{q} = \frac{M}{Q} \tag{3.3}$$

Its inverse is the specific airflow rate or air change rate, n.

If the room or ventilated zone has a defined air exhaust, Sandberg (1984) has shown that the nominal time constant is equal to the mean age of air at this exhaust:

$$\tau_n = \bar{\tau}_e \tag{3.4}$$

Air exchange efficiency

This efficiency expresses how the fresh air is distributed in the room. The time, τ_a, required on average to replace the air present in the space is twice the room mean age of air (Sandberg and Sjöberg, 1983):

$$\tau_a = 2\langle \tau \rangle \tag{3.5}$$

At a given flow rate and zone volume, the shortest time required to replace the air within the space is given by the nominal time constant. Therefore, the air exchange efficiency, η_a, is calculated by:

$$\eta_a = \frac{\tau_n}{2\langle \tau \rangle} \tag{3.6}$$

The air exchange efficiency is equal to one for piston-type ventilation, where the exhaust is reached at a time corresponding exactly to the nominal time

Piston ventilation	Full mixing	Shortcuts and dead zones
$0.5 < \eta_a \leq 1$	$\eta_a \approx 0.5$	$\eta_a \leq 0.5$

Figure 3.1 *Ventilation modes with typical airflow patterns and air change efficiencies*

Source: Roulet, 2004.

Table 3.1 *Nominal time constant and room mean age of the air corresponding to the probability curves shown in Figures 3.2 and 3.3*

Air exchange efficiency	η_a	25%	50%	66%	80%	90%	99%
Room mean age of the air	$\langle \tau \rangle$	1.44	0.99	0.76	0.62	0.55	0.50

constant. In rooms with complete mixing, the room mean age of air equals the nominal time constant, and the air exchange efficiency is 50 per cent. Short-circuiting of air will also leave dead zones in the room, giving rise to an efficiency that is lower than 0.5 (see Figure 3.1).

Table 3.1 shows the air exchange efficiency and the corresponding room mean age of the air, assuming that the nominal time constant or the mean age at exhaust is one hour.

Some typical probability density curves of the age of the air at the exhaust are illustrated in Figure 3.2, the air exchange efficiency being used as parameter. If the air is displaced like a piston, all air particles reach a given location in the room at the same time, and they reach the exhaust at a time

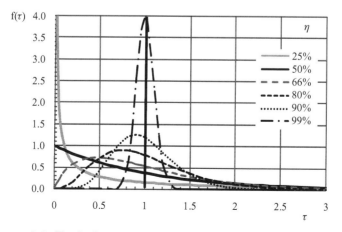

Figure 3.2 *Typical probability density curves for the age of the air*
Note: The parameter is the air exchange efficiency.

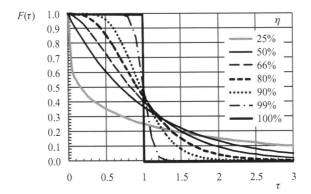

Figure 3.3 *Typical probability curves for the age of the air*

Note: These theoretical curves are for illustration. Some of them, in particular at very high efficiency, are not likely to be found in practice.

corresponding to the nominal time constant. The air exchange efficiency in this very theoretical case is 100 per cent. At 99 per cent air exchange efficiency, there is already some spreading of the ages around an average still equal to the nominal time constant. When probability density function spreads out and the probability function smoothens, the mean age at exhaust remains the same, but there are more young air particles and more aged ones. In addition, the most probable age (the time of the maximum of the curve) is reduced. At 66 per cent air change efficiency, this most probable age is already half the nominal time constant. At 50 per cent efficiency or complete mixing, the probability density of the age of the air at exhaust is an exponential: there are more young particles than old ones reaching the exhaust. With this distribution, the most probable age is zero, but the mean age is still equal to the nominal time constant. The last curve, with 25 per cent efficiency, represents a situation with a shortcut.

Figure 3.3 shows the corresponding probability curves for the age of the air at exhaust, still with one hour as nominal time constant. This function shows the percentage of particles at exhaust that are older than the value given on the ordinate. At 100 per cent efficiency, all particles are exactly one hour (one time constant) old. When the efficiency decreases, the function progressively changes to become an exponential at 50 per cent efficiency or complete mixing. When the efficiency decreases, there are less and less young particles and more and more old ones at exhaust. At 25 per cent efficiency, which is very poor, 10 per cent of the volume of the air at exhaust is older than three time constants!

Measurement method

The air entering the room is marked with a gas (the tracer gas), and the concentration of that tracer gas is monitored at the location of interest. This assumes that the tracer gas behaves the same as the air: no adsorption and same buoyancy, which is the case if the tracer concentration is small. It can be readily understood that if the air is marked at the inlet by a short pulse of tracer gas, and if the tracer molecules follow the air molecules, they will arrive at a given location at the same time as the air molecules. The time spent between injection and the detection of most tracer gas molecules by the analyser is the age of the tracer, hence the age of the air at the air sampling location. The pulse technique is, however, not the only one and the following three strategies can be used:

- Step-down – a uniform concentration of tracer is achieved at the beginning of the test, when the injection is stopped;
- Step-up – the tracer is injected at the air inlet at a constant rate from the starting time throughout the test;
- Pulse – a short pulse of tracer is released in the air inlet at the starting time.

The probability functions and the local mean ages at any point, r, can be calculated from the time history of the net tracer concentration, $C_r(t)$, which is the measured concentration minus the background concentration. It was shown, however, that for rooms with a single air inlet and a single air outlet, the step-up method should be preferred, since it is the easiest to perform and gives the best accuracy (Roulet and Cretton, 1992).

For the step-up technique, the tracer gas is injected at a constant rate into the supply air in the outside air duct, starting at a known time t_0. It should be ensured that the tracer and the air are fully mixed in the supply duct to produce a steady concentration, C_3, at the air inlet. If C_3 cannot be measured, the equilibrium concentration within the enclosure, C_4, may be used instead. The notations for air sampling refer to Figure 2.5.

Tracer gas concentration at the locations where the age of air is looked for is recorded. The sampling time interval should be short enough to record the transient evolution of the concentration. It should then be much shorter than the expected age of air. A good value is one tenth of the nominal time constant, which can be estimated by dividing the ventilated volume by the design airflow rate, or better, by the actual airflow rate if it is known.

One important location is in the exhaust duct, where $C_e = C_6$ is measured. Recording the evolution of the tracer gas concentration at this location provides both the actual nominal time constant and the mean age of air in the ventilated space. Injection rate is maintained constant until a steady state is obtained. Depending on the air change efficiency, this may take up to four time constants. An example of such a record is given in Figure 3.4.

When the concentration stabilizes at a value noted, C_∞, the step-up experiment is ended. However, it is recommended to continue recording the concentration after having stopped the tracer gas injection, since this will provide a second measurement of the age of the air, using the decay method. For this purpose, the time when injection is closed should be noted, since this time is the starting time of the decay experiment.

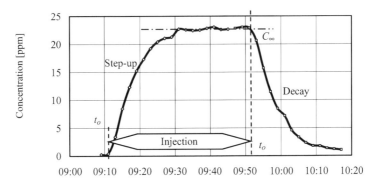

Figure 3.4 *Record of tracer gas concentration in the exhaust duct during the measurement of the age of air*

To interpret the recorded tracer gas concentrations and obtain the age of air, the background (or supply) concentration should first be subtracted from all measurements, and the elapsed time should be calculated by subtracting the starting time from all time values. In the following formulae, the net tracer gas concentration, C_r, is the difference between the concentration measured at location r and the concentration of this gas in the outdoor air.

The probability function of the age of air can be calculated from the concentration ratio:

$$\text{Step-up } F(\tau) = 1 - \frac{C(t - t_0)}{C_\infty} \qquad \text{Decay } F(\tau) = \frac{C(t - t_0)}{C(t_0)} \qquad (3.7)$$

Figure 3.5 shows the concentration ratio calculated from the recorded concentration illustrated in Figure 3.4.

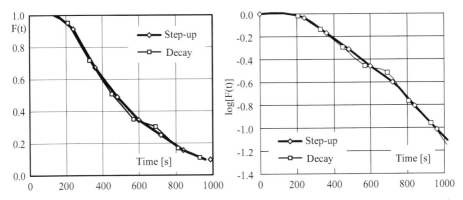

Figure 3.5 *Probability functions of the age of air, calculated from the recorded concentration illustrated in Figure 3.4*

Note: Left is a linear scale and right is a logarithmic scale, showing an exponential decay after 700 s.

The local mean age of air at any location is the integral (or zero moment) of the probability distribution:

$$\tau_r = \mu_0 = \int_0^\infty F_r(t)\, dt \tag{3.8}$$

The first moment of the probability distribution is, by definition:

$$\mu_1 = \int_0^\infty t F_r(t)\, dt \tag{3.9}$$

If there is only one single exhaust, the room mean age of air can be deduced from tracer concentration measurements in the exhaust duct, $C_e(t)$:

$$\langle \tau \rangle = \left(\frac{\mu_1}{\mu_0} \right)_e = \frac{\displaystyle\int_0^\infty t F_e(t)\, dt}{\displaystyle\int_0^\infty F_e(t)\, dt} \tag{3.10}$$

In this case, the nominal time constant of the ventilated space, τ_n, which is the ratio of the space volume and the volumetric airflow rate, is equal to the mean age of air at the exhaust:

$$\tau_n = \bar{\tau}_e = \int_0^\infty F_e(t)\, dt \tag{3.11}$$

Therefore, the air exchange efficiency, η_a, can be assessed directly by measuring the evolution of the concentration at the exhaust:

$$\eta_a = \frac{\tau_n}{2\langle \tau \rangle} = \frac{\tau_e}{2\langle \tau \rangle} = \left(\frac{\mu_0^2}{2\mu_1} \right)_e \tag{3.12}$$

Practical interpretation of the concentration records

In practice, the moments in the above formulae are calculated numerically, on the basis of discrete recorded values of the concentration and time. A simple way to calculate these moments uses the trapezium integration method, with the general formulation of:

$$\int_0^{t_N} f(t)\, dt \cong \sum_{j=0}^{N-1} \frac{1}{2}(f_j + f_{j+1})\Delta t \tag{3.13}$$

where f_j is for $f(t_j)$ and Δt for $t_{j+1} - t_j$.

Approximating the variation of the concentration during each time step by a straight line, the two integrals defined above can be estimated by summing finite elements:

$$\mu_0 = \int_0^\infty F_e(t)\, dt = \left(\frac{F_0 + F_N}{2} + \sum_{j=1}^{N-1} F_j \right) \Delta t + \varepsilon_0(N, \tau_d) \tag{3.14}$$

$$\mu_1 = \int_0^\infty t F_e(t)\, dt = \left(\frac{N F_N}{2} + \sum_{j=1}^{N-1} j F_j \right) \Delta t^2 + \varepsilon_1(N, \tau_d) \tag{3.15}$$

where:

F_j is the probability distribution at time $t = j \Delta t$,

Step-up case $F_j = 1 - \dfrac{C(t_0 + j\Delta t)}{C(\infty)}$ Decay case $F_j = \dfrac{C(t_0 + j\Delta t)}{C(t_0)}$

$$(3.16)$$

N is the last measurement integrated using the trapezium method,
$\varepsilon_n(N, \tau_d)$ is the rest of the integral, evaluated using an exponential fit on the last
           measurements (see below).

The number of measurements, N, could be large enough to ensure that the
sum of the terms for $j > N$ are negligible, or, in other words, that C_N is very
close to the steady-state value. In this case, the remaining parts, $\varepsilon_n(N, \tau_d)$,
are negligible. In practice, however, the measurement can be stopped before
reaching the steady state. In this case, the tail in the integral of the moments
is not measured, but is estimated. As shown in Figure 3.5, this tail is, in
most cases, exponential of the form:

Step-up: $C(t) = C_\infty (1 - e^{-t/\tau_c})$
Decay: $C(t) = C(0)e^{-t/\tau_c}$

$$(3.17)$$

Therefore, for time larger than $t_N = N \Delta t$, it can be assumed that:

$$F(t > t_N) = F_N \cdot \exp\left(\frac{t_N - t}{\tau_d}\right) \qquad (3.18)$$

where τ_d is a time constant determined by a fit on the last measurements in
the exponential part. In this case, the remaining part, $\varepsilon_n(N, \tau_d)$, of the
moments are:

$$\varepsilon_0(N, \tau_d) = \int_{t_N}^{\infty} F_e(t)\, dt = F_N \int_{t_N}^{\infty} \exp\left(\frac{t_N - t}{\tau_d}\right) dt = F_N \tau_d$$

$$\varepsilon_1(N, \tau_d) = \int_{t_N}^{\infty} t F_e(t)\, dt = F_N \tau_d (t_N + \tau_d)$$

$$(3.19)$$

The time required for reaching an exponential decay depends not only on the
nominal time constant of the room, but also on the ventilation system. The
decay will be exponential from the beginning of the test where complete
mixing occurs. In case of displacement ventilation, the decay should be very
sharp after a time equal to the age of air.

Error analysis

Using, *mutatis mutandis* Equation 2.22, the confidence interval of $F(\tau)$ is:

$$\lfloor F_j - \delta F_j; F_j + \delta F_j \rfloor$$

with

$$\delta F_j = T(P,\infty)\sqrt{\left(\frac{\delta C_e(j\Delta t + t_0)}{C_e(j\Delta t + t_0)}\right)^2 + \left(\frac{C_e(j\Delta t + t_0)}{C_e^2(\infty)}\delta C_e(\infty)\right)^2} \qquad (3.20)$$

Then, the confidence intervals of the moments are:

$$\delta\mu_0 = T(P,\infty)\sqrt{f_{\mu_0}} \qquad \text{and} \qquad \delta\mu_1 = T(P,\infty)\sqrt{f_{\mu_1}} \qquad (3.21)$$

with

$$f_{\mu_0} = \left(\frac{\Delta t}{2}\right)^2 \left((\delta F_0)^2 + (\delta F_N)^2\right) + (\Delta t)^2 \sum_{j=1}^{N-1}\delta F_j^2$$

$$+ \left(\frac{F_0 + F_N}{2} + \sum_{j=1}^{N-1}F_j\right)^2 (\delta\Delta t)^2 + (\delta\varepsilon_0)^2 \qquad (3.22)$$

and

$$f_{\mu_1} = \left(\frac{\Delta t}{2}\right)^2 \left((\delta F_0)^2 + (\delta F_N)^2\right) + (\Delta t)^2 \sum_{j=1}^{N-1}j^2\delta F_j^2$$

$$+ \left(\frac{NF_N}{2} + \sum_{j=1}^{N-1}jF_j\right)^2 (\delta\Delta t)^2 + (\delta\varepsilon_1)^2 \qquad (3.23)$$

in which

$$\delta\varepsilon_0 = T(P,\infty)\sqrt{\tau_d^2\delta F_N^2 + F_N^2(\delta\tau_d)^2} \qquad (3.24)$$

and

$$\delta\varepsilon_1 = T(P,\infty)\sqrt{(t_N + \tau_d)^2\tau_d^2\delta F_N^2 + (t_N + 2\tau_d)^2 F_N^2(\delta\tau_d)^2} \qquad (3.25)$$

where $\delta\tau_d$ is the confidence interval resulting from the exponential fit.
Finally we get:

$$\delta\tau_n = \delta\tau_e = \delta\mu_0(F_e) \qquad (3.26)$$

$$\delta\langle\tau\rangle = \left(\frac{1}{(\mu_0(F_e))^2}\right)\sqrt{(\mu_1(F_e))^2(\delta\mu_0(F_e))^2 + (\delta\mu_1(F_e))^2} \qquad (3.27)$$

and

$$\delta\eta_a = \frac{\delta\mu_1(F_e)}{(2\mu_0(F_e))^2} \qquad (3.28)$$

Example of application

The ventilation system of a 60-seat conference room was retrofitted to improve indoor air quality. The old, mixing-type installation was replaced by

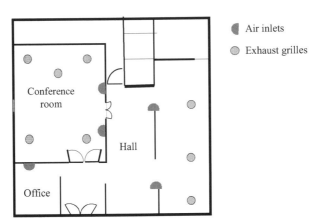

Figure 3.6 *Arrangement of the conference room and of its surroundings*
Source: Roulet *et al.*, 1998.

a displacement ventilation system, as shown in Figure 3.6. The conference room is 8 m by 10 m wide and 3 m high. It is completely embedded in an old, massive building. Its walls, floor or ceiling have no contact with the outdoor environment. It has no windows, but leaky entrance doors leading to a hall.

The mechanical displacement ventilation system includes two low-velocity air inlets, 1 m high, put at the floor level against one wall of the room. The air, slightly colder than the room temperature, is introduced at low speed close to the ground through three inlets. This cold air spreads on the ground like a lake, and should go up faster where there are heat sources like occupants. This asymmetric disposition does not allow a uniform distribution of fresh air in the room, but was first adopted for practical reasons, the building layout not allowing an optimum location of air inlets and outlets. Five exhaust grilles are located in the ceiling. Since the building owner was interested in assessing the actual performance of the new system, age of air measurements were performed in order to check if the airflow pattern in the room was as expected.

First, age of air measurements were performed in the room as it was. It was found that the air was poorly distributed in the room because of the asymmetrical disposition of inlet grilles. In addition the doors were found to be leaky, and much air from the ventilation system was leaving the room quickly after entering it, thus reducing the purging effect. The indoor air quality was, however, good, since the ventilation rate was very high (eight minutes nominal time constant, or 15 outdoor air change per hour, or 60 m³/h per occupant at full occupancy!).

On the basis of these results, improvements were brought to the system. The leakages in the room envelope were sealed, a door was added, a new inlet was added on the wall opposite to the existing inlets, and an exhaust grille was added in the ceiling as shown in Figure 3.7.

A second measurement campaign was performed, showing a significant improvement of the ventilation efficiency (see Figure 3.8). The air change

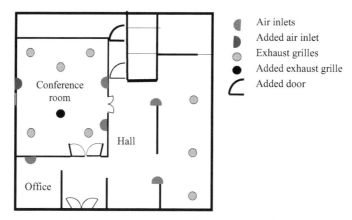

Figure 3.7 *Arrangement of the conference room after improvement*
Source: Roulet *et al.*, 1998.

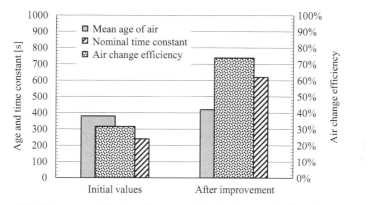

Figure 3.8 *Room ventilation characteristics before and after improvement*
Source: Roulet *et al.*, 1998.

efficiency was doubled, and the mean age of air was maintained despite a reduction of the ventilation rate – and of energy use – by a factor two.

Mapping the age of the air in rooms

This chapter demonstrates how to assess the age of air at some location and on the average in a room. It may nevertheless be interesting to map this quantity in a room, in order to check, for example, if the occupants have the best possible air, or to look for dead zones. Davidson and Olsson (1987) have already generated such maps using computer codes, and some qualitative representations have been drawn from measurements (Valton, 1989). Since the measurement of the age of the air at a given location is not straightforward, takes time and has its cost, the theory of experimental design (Box *et al.*, 1978) may help in providing a maximum of information through a minimum of measurements.

Minimum number of measurements

A map of any scalar variable, v, in a three-dimensional room is in principle obtained by measuring the variable at each node of a network and interpolating between these nodes. Such measurements are, however, very expensive and may be unfeasible. If only five values are taken on each axis, at least 125 measurements are required, meaning analysis of the air every few minutes at 125 locations over a couple of hours. Therefore, it makes sense to apply a method that needs a minimum number of measurements points. This minimum number depends on the objective of the mapping experiment, or more precisely on the required mapping details. Since the interpolation between measurement points needs a model, the mapping network indeed depends on the empirical model chosen to represent the map of the variable, v.

Any infinitely derivable function (as v is assumed to be) can be developed in a Taylor series around a given point. This gives a polynomial, which can be approximated by its $k + 1$ first terms, k being the degree of the polynomial. In the following, models of degree 1 and 2 are considered. If a linear model is adopted (degree 1), such as:

$$v = a + \sum_i b_i x_i \qquad (3.29)$$

where x_i are the three coordinates of the measured point, only four measurements are needed to obtain a set of coefficients (a, b_i). If more measurements are made, the coefficients may be obtained by a least square fit procedure provided there is no (or negligible) uncertainty on the coordinates. If their coordinates differ for the other points, these supplementary measurement points give information on the validity of the model used.

If the linear model does not appear to be valid, higher degree models may be used. For example, a quadratic model:

$$v = a + \sum_i b_i x_i + \sum_{i \neq j} b_{ij} x_i x_j + \sum_i b_{ii} x_i^2 \qquad (3.30)$$

that contains ten coefficients, can be chosen. Such a model may already fit many practical situations and present minimal and maximal value(s). To determine its coefficients, measurements taken at ten locations are the minimum necessary.

An intermediate model is the interactions model:

$$v = a + \sum_i b_i x_i + \sum_{i \neq j} b_{ij} x_i x_j \qquad (3.31)$$

Table 3.2 *Minimum number of measurements needed to obtain the coefficients of a kth degree polynomial empirical model representing a variable in a two- and three-dimensional space*

Model dimensions	Linear	Interaction	Quadratic	Cubic	4th degree
2	3	4	6	10	15
3	4	7	10	20	35

for which seven coefficients must be determined. Table 3.2 summarizes the minimum number of measurements needed.

Location of the measurement points

An important issue is the appropriate location of measurement points. The set of measurement points is called an experimental design. There are many possible experimental designs, but they do not give the expected results with the same accuracy. For example, it is obvious that to fit a linear model in one dimension only (the straight line modelled by $y = ax + b$), the location of the two measurement points (the minimum number) that gives the best accuracy for the coefficients a and b is at the extremities of the experimental domain, i.e. at the minimum and maximum possible values of the variable, x.

For more sophisticated models or in a larger number of dimensions, the locations of the best sampling points are not so obvious. However, several tools exist for planning such experiments, which are found in the literature (Fedorov, 1972; Box *et al.*, 1978; Bandemer and Bellmann, 1979; Feneuille *et al.*, 1983; Aeschlimann *et al.*, 1986) and are applied below.

In experiments to determine the age of the air, points close to walls do not represent the inner volume, and the sampling points should not be located too close to walls or in the corners of the room. In the following, the 'room' or the 'experimental domain' is a volume that is smaller than the actual measured space by about 20 per cent in each direction.

Let us take a coordinate system in such a rectangular volume using as the unit, for each direction, the half-length of that domain in that direction. Three numbers, included in the interval $[-1, +1]$, locate any point in the 'room'.

The experimental design can be represented by a rectangular matrix with three columns (one for each coordinate) and as many lines as measurement points. A general condition is that in order to obtain the coefficients of a polynomial of degree k, each of the variables x, y and z shall take at least $k + 1$ values in the experimental design, which should have at least $k + 1$ levels on each axis.

The criteria described below are used to establish the most efficient design.

Criteria for location of the measurement points

The model matrix M

First, let us look at the method used to obtain these coefficients. For each point, the model is applied, replacing the x_i by their values given by the experimental design. A system of equations (one equation for each location) is obtained this way, which can be written in a matrix notation:

$$V = MA \tag{3.32}$$

where:

V $(v_1, v_2, \ldots, v_n)$ is a vector containing the measured quantities at the n locations.

M is a matrix, each line of which corresponding to one location. Its first column is filled with ones and corresponds to a constant in the model. The next three columns may contain the coordinates of the locations if the model contains linear terms. The next three columns may contain the products of these coordinates two by two (for example, x_1x_2, x_1x_3, x_2x_3) in case of interaction terms and, for a quadratic model, the next three columns contain the squares of the coordinates. Other models will produce other matrices.

A is a vector containing the coefficients (e.g. a, b_i, b_{ij} $(i \neq j)$ and b_{ii}) of the model.

In the general case, **M** is rectangular and the least square fit procedure is used:

$$A = (M^T M)^{-1} M' V \tag{3.33}$$

where M^T is the transposed matrix of **M**. This equation is also valid if **M** is a square matrix, but reduces to the simpler equation:

$$A = M^{-1} V \tag{3.34}$$

In any case, a matrix should be inverted and the determinant of this matrix should not be zero! Since this determinant can be calculated *before* making the measurements, it is a first criterion for the choice of the experimental design: it should be significantly different from zero.

Variance of the calculated response

If the coefficients are known, an estimate v_e of the value of the variable v can be obtained at each location in the enclosure:

$$v_e = A^T r \tag{3.35}$$

where r is the vector $(1, x_1, x_2, x_3)$.

If σ^2 is the experimental variance of the measured variable v, the variance $\sigma^2(v_e)$ of the estimated variable is:

$$\sigma^2(v_e) = \boldsymbol{r}^T(\boldsymbol{M}^T\boldsymbol{M})^{-1}\boldsymbol{r}\sigma^2 \tag{3.36}$$

A variance function can be defined:

$$VF = \frac{N}{\sigma^2}\sigma^2(v_e) = N\boldsymbol{r}^T(\boldsymbol{M}^T\boldsymbol{M})^{-1}\boldsymbol{r} \tag{3.37}$$

where N is the number of measurements. VF depends on the experimental design ($\boldsymbol{M}$ and N) and on the location $\boldsymbol{r}$ and can hence be calculated *before* doing any measurement.

If VF depends only on the distance to the origin (or the modulus of $\boldsymbol{r}$), the experimental design is said to be isovariant by rotation. If VF is a constant within the experimental domain, the design gives a uniform accuracy. A good experimental design should have a small variance function, as constant as possible.

If $(\boldsymbol{M}^T\boldsymbol{M})^{-1}$ is diagonal, the design is orthogonal. In this case, the variance function is minimum.

Condition of the model matrix

The condition number of the matrix $\boldsymbol{M}$ plays an important role on the upper bound of the relative errors on the result (see Chapter 7, 'Upper bound of the errors'). This condition multiplies the experimental errors and transfers these errors into the result $\boldsymbol{A}$. It should therefore be as small as possible. This number depends on the experimental design and on the model chosen but does not depend on the results of the measurements. Hence it can be calculated *before* doing any measurement and constitutes one more criterion, which is relatively easy to compute, for the choice of the best experimental design. This is a much better criterion than the determinant of $\boldsymbol{M}^T\boldsymbol{M}$.

Expendability of the experimental design

It may be interesting that the measurements performed to obtain the co-efficients of a first-degree polynomial are not lost and could be used with other measurements to expand the polynomial to a higher degree. Some designs are expandable that way, some others are not.

Examples of experimental designs

Several experimental designs for mapping parallelepiped volumes or a rectangular area were examined (Roulet *et al.*, 1991). Linear, interaction and quadratic models were tested. Several of these designs were found to be unusable (singular matrix or too large a condition number for the quadratic model). Only good examples are given below.

As mentioned above, the experimental domain is about 20 per cent smaller than the measured space, samples of air being taken at least 0.1 times the characteristic enclosure dimension from the walls.

Factorial designs

A factorial design for k dimensions and l levels is obtained by dividing the experimental domain (for example, the interval $[-1, 1]$) on each axis into l equidistant levels. The complete factorial design contains all the points obtained by the l^k combination of the l possible values of the k coordinates.

The number of points in a full factorial design is hence l^k. If l and k are greater than 2, the full factorial designs often have many more points than the minimum required, and are therefore seldom used. However, partial factorial designs can be obtained by selecting the necessary number of measurement points from the full design. Some examples are given below.

The 2-D, two-level full factorial design (see Table 3.3) is optimal for a linear model, providing the coefficients of that model with the best accuracy. If, for economical reasons, one point is omitted, the confidence intervals of the coefficients are twice that based on four measurement points.

Table 3.3 *2-D, two-level full factorial design*

No	x	y
1	−1	−1
2	1	−1
3	−1	1
4	1	1

It is very important to note that the frequently used design consisting of changing one variable at a time (see Table 3.4) is less accurate than the former.

Table 3.4 *2-D design changing one variable at a time*

No	x	y
1	1	0
2	0	1
3	−1	0
4	0	−1

Adding a fifth point at the centre $(0,0)$ of the 2-D, two-level full factorial design allows assessing the coefficient of the interaction term b_{12}, without loss of accuracy. The following two points:

No	x	y
6	−1	0
7	1	0

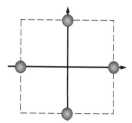

Figure 3.9 *Minimum design for a 2-D quadratic model*

can be added to obtain a minimum design for a quadratic model, which has a condition number of 6.3 (see Figure 3.9).

The 2-D full factorial design with three levels shown in Table 3.5 has a better condition number (4.4) for a quadratic model.

Table 3.5 *2-D full factorial design with three levels*

No	x	y	No	x	y
1	−1	−1	6	1	0
2	0	−1	7	−1	1
3	1	−1	8	0	1
4	−1	0	9	1	1
5	0	0			

3-D designs

In three dimensions, the four point design of Table 3.6 is best for a linear model.

Table 3.6 *Minimum 3-D design for assessing the coefficients of a linear model*

No	x	y	z
1	−1	−1	1
2	1	−1	−1
3	−1	1	−1
4	1	1	1

It can be expanded to a full factorial design (see Table 3.7), which is appropriate for an interaction model.

Also at three dimensions, the star design shown in Table 3.8 is less accurate and requires more work than the minimum design of Table 3.6.

However, combining the centred star design with the full factorial design of Table 3.7 gives a so-called composite centred design, suitable for a quadratic model, having a condition number of 4.4. If fewer points are wanted,

Table 3.7 *Full factorial design for assessing the coefficients of a linear model with interactions*

No	x	y	z
1	−1	−1	−1
2	1	−1	−1
3	−1	1	−1
4	1	1	−1
5	−1	−1	1
6	1	−1	1
7	−1	1	1
8	1	1	1

Table 3.8 *3-D centred star design*

No	x	y	z
9	1	0	0
10	0	1	0
11	0	0	1
12	−1	0	0
13	0	−1	0
14	0	0	−1
15	0	0	0

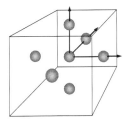

the points 8, 5 and 2 can be deleted (in that order) giving finally a design having 12 points and a condition number of 4.8. Finally, deleting two more points (3 and 15) gives the design C3, which has six points in the centre of the faces and four points at opposite corners (see Figure 3.10).

The condition number of $M^T M$ calculated using the absolute value norm $|M^T M|$ for these designs and three models is given in Table 3.9.

There are numerous other possibilities that can be imagined or found in the literature.

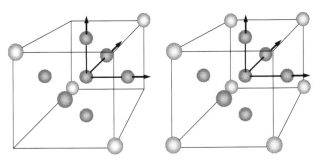

Figure 3.10 *Experimental designs C3 (left) and composite centred (right)*

Table 3.9 *Condition number of $\mathbf{M}^T\mathbf{M}$ for some experimental designs and three models*

Experimental design	Number of points	Quadratic model	Interactions model	Linear model
2-D Designs				
2-level part factorial	3	–	–	2.0
2-level full factorial	4	–	–	1.0
Centred 2-level factorial	5	–	1.0	1.0
Minimum for quadratic	6	6.3	1.0	1.0
3-level full factorial	9	4.4	1.0	1.0
3-D Designs				
2-level half factorial	4	–	–	1.0
2-level full factorial	8	–	1.0	1.0
C3	10	4.3	3.2	1.0
Composite centred	15	4.4	1.0	1.0

Example of application

The age of air was mapped in the conference room after the improvement, first in the empty room, and then with ten occupants sitting around a U-shaped table. These measurements were performed at the head level of sitting persons, using the nine-sampling-point full factorial design shown in Table 3.5. The results are shown in Figure 3.11. In the unoccupied room, the air is older at the middle left, where there is only one air inlet. When the room is occupied, the air is younger in the middle of the room, where the occupants are.

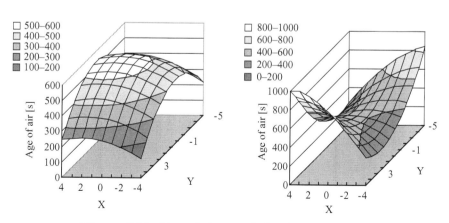

Figure 3.11 *Map of the age of the air at head level*
Note: Left is the unoccupied room, and right is the room occupied by ten persons sitting around a conference table.

4

Airtightness

Why check airtightness?

Controlled airflows, having adequate flow rate and passing at the appropriate locations are essential for good indoor air quality. Leakage, allowing the uncontrolled air to follow inappropriate paths, should therefore be reduced as much as possible. This requires an airtight building envelope and airtight ductwork: building and ductwork airtightness is a prerequisite for efficient natural or mechanical ventilation. Envelope leakage is not an appropriate way for airing buildings.

In buildings with hybrid ventilation, the indoor air quality is controlled partly by a mechanical ventilation system and partly by a natural ventilation design. The share between these two systems could be seasonal (natural ventilation in mild seasons and mechanical ventilation with heat recovery during hot or cold seasons) or spatial: small rooms with an external wall with natural ventilation, and large rooms or rooms located inside the building with mechanical ventilation. In any case, both ventilation systems should be controlled.

While Figure 0.5 shows the exfiltration ratios (i.e. the part of the supplied air leaving the building by another way than the exhaust duct) for several buildings, Figure 4.1 shows the infiltration ratios (i.e. the part of outdoor air that is not supplied by the mechanical ventilation unit) measured in 11 spaces that are equipped with full mechanical (not hybrid) ventilation. Out of these ten spaces, seven have an infiltration rate significantly different from zero, and in two of them more than 30 per cent of the outdoor air is not controlled!

Exfiltration has a negative effect on heat recovery, since the heat in the air leaving the building through leakage cannot be recovered (see Chapter 5, 'Effect of leakages and shortcuts on heat recovery'). In cold climates, warm and humid indoor air going through the external envelope through leaks encounters increasingly colder surfaces. The water vapour of this air eventually condenses on the coldest surfaces within the cracks, thus creating dramatic condensation problems at leakage locations. Infiltration has a negative effect on indoor air quality, since infiltrated air is neither filtered, dried, cooled nor heated.

A survey performed by Carrie *et al.* (1997) in France and Belgium has shown that, on the average, 40 per cent of the supplied air is lost through ductwork leakage before reaching the user. This obviously reduces the effective

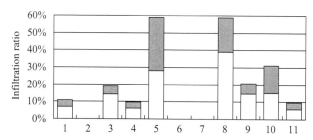

Figure 4.1 *Measured part of outdoor air that is not supplied by the system in mechanically ventilated buildings, shown with uncertainty band*

ventilation rate. Maintaining air quality requires an increase in supply air, leading to energy wastage.

Checking the airtightness of a building envelope or a duct network should therefore be performed at each commissioning of a building or ventilation system.

Measurement methods

The airtightness of the envelope of the measured object is in fact expressed by its permeability to air, which is the relationship between the leakage airflow rate and the pressure differential through the envelope of the object. This relationship can be either expressed by a mathematical expression (see Equations 4.1 and 4.3) or by an equivalent leakage area (see Equation 4.19).

One internationally standardized way to assess this permeability requires the maintenance of a pressure differential between the interior and the exterior of the object with a fan, and the measurement of the airflow rate needed to maintain this pressure differential (ISO, 1998). Other, simpler but less accurate methods applicable to buildings are described below in 'The stack effect method'. They use the stack effect to create the pressure differential and openings and the location of the neutral level (the level where the indoor–outdoor pressure differential is zero) to estimate the leakage area.

The fan pressurization method

Before measurements are taken, all purpose-installed openings (doors, windows, ventilators, etc.) should be closed, and the mechanical ventilation system switched off (unless used for pressurization). The ventilation grilles should either be sealed or all ventilation dampers should be closed. It may be necessary to seal chimneys and flues but these can be sealed later as part of the test if desired. Clean up open fires to avoid dispersion of ash in the rooms.

The pressure differential is created by a fan installed in an opening of the envelope of the object, or by the ventilation fans themselves. It is measured with a sensitive manometer (range 0–100 Pa (Pascals) or 10 mm water column) and the airflow rate through the fan is measured using any of the following methods:

- The airflow rate through a fan depends on the pressure differential and its rotation speed. Measuring these two quantities allows assessment of the airflow rate from the characteristic curve of the fan. Blower doors use this method.
- A suitable airflow meter such a nozzle or a sharp-edged orifice is installed in the airflow circuit (see Figure 4.5).
- The tracer gas dilution technique, as described in Chapter 2, 'Tracer gas dilution'.

These measurements are repeated for several pressure differentials, ranging from a few Pascals to about 60 Pa, or even more for some cases.

The minimum pressure is limited by the noise of the pressure differential, for example, the random pressure variations resulting from wind and stack effect. Therefore, measurements should be performed when there is no wind and the minimum pressure differential is in practice twice the natural pressure differential. The maximum pressure is limited by the resistance of the object by practical limits such as fan airflow rate combined with the object's leakage. Note that 100 Pa is a pressure that can result from 40 km/h wind velocity.

Since fan pressurization is subject to the disturbing influence of natural pressure fluctuations created by the wind, most measurements are made at pressure differentials far above those created by natural forces. This may lead to inaccuracy if the results are extrapolated to lower pressure differentials.

Two general models are used to characterize air permeability. The power law, fully empirical, reflects the fact that leakage is a combination of various cracks and openings that may be arranged in parallel and series network:

$$q = C\Delta p^n \tag{4.1}$$

where:

q is the volume airflow rate through the leakage site (m^3/s);
Δp is the pressure difference across the leakage site (Pa);
n is the flow exponent $(0.5 < n < 1)$;
C is the airflow coefficient $(m^3\,s^{-1}\,Pa^{-n})$.

Since the airflow may be either laminar or turbulent, and the airflow rate is proportional to the pressure differential in laminar flows and to its square root in turbulent flows, Etheridge proposes a quadratic law, that expresses that the flow is a mix of laminar and turbulent flow arranged in parallel (Etheridge and Sandberg, 1996):

$$\Delta p = aq^2 + bq \tag{4.2}$$

where:

a and b are coefficients representing respectively the turbulent and laminar parts of the quadratic law $(Pa\,s/m^3$ and $Pa\,s^2/m^6)$.

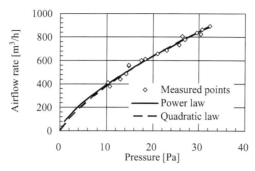

Figure 4.2 *Airflow rates and pressure differences as measured in a real test, together with power law and quadratic fits*

Inverting this relationship gives the airflow rate resulting from a pressure differential:

$$q = \frac{-b + \sqrt{b^2 - 4a\Delta p}}{2a} \tag{4.3}$$

By fitting one of these two models on the measured points on a $[q; \Delta p]$ diagram, the coefficients and the exponent of either of the above models can be assessed (see Figure 4.2).

Reductive sealing

Global permeability measurement does not provide information on the location of leakage. Reductive sealing aims to quantify the proportion of the total air leakage that is attributable to different components or groups of components. One method to locate leakage is to seal components, areas, rooms or zones suspected of leakage and to perform a new pressurization test. The difference is the leakage of the sealed leaks.

The fan pressurization equipment is set up and the dwelling prepared, in the usual way (see 'The fan pressurization method', above). An air leakage test is then carried out, either pressurizing or depressurizing. A component or group of components is then chosen (for example, all chimneys and flues) and sealed with self-adhesive tape, polyethylene sheet, inflatable bladder or modelling clay, as appropriate. It is important that care is taken to make a good seal on all components. For example, special attention should be given to corners and around the fasteners.

The air leakage test is then repeated, the difference between this test and the first test being a measure of the air leakage attributable to the component or group of components that were sealed.

Further components can then be chosen and the process continued. The difference of leakage flow rate at each pressure between two successive tests is the leakage of the components sealed between these tests. Leakage coefficients for each group of sealed components can be calculated from these leakage flow rates at each pressure using the method described in section 'Determining the leakage coefficients', below.

Components that can usually be sealed include: chimneys, flues; ventilation openings, external doors (other than the one used to mount the pressurization fan), wooden ground floors or roof, cracks between walls and floors, pipe and cable entry or exit points, and any other obvious cracks and openings (for example, gaps between a window frame and the wall into which it is fixed).

The last air leakage test in this process gives a measure of the background air leakage, i.e. the remaining air leakage not sealed during the previous tests. It is quite common for this to constitute more than half of the overall air leakage, even where all of the most obvious air leakage paths have been sealed.

When all the components required have been sealed and air leakage tests carried out, the pressurization fan can be reversed and the air leakage tests repeated as the components are progressively unsealed in the reverse order to that in which they were sealed.

Best results are usually obtained by sealing groups of components (for example, all openable windows in the dwelling) because the leakage through an individual component (for example, a single openable window) can be too small for the pressurization fan to resolve.

Multi-zone fan pressurization method

The technique can be extended to multi-zone pressurization, aiming to assess the air permeability of not only the envelope of an object (for example, a building) but also of its internal partitions (Fürbringer and Roulet, 1991). The building zones are represented by nodes of a network linked by partitions. One of the zones is the outdoor air. To assess the coefficients of all partitions requires the measurement of many inter-zone airflow rates and pressure differentials. An appropriate design of the experiment aiming to assess the required coefficients, and only these, will considerably reduce the work required.

For example, two fans and a control system allow assessment of the leakages of many parts of a building. To avoid the tedious work of sealing with plastic foil all building parts that should not be measured, the airflows through these parts are inhibited by maintaining a zero pressure difference across them.

The measuring fan with its flowmeter equipment is installed in a wall or door of the room containing the element to be measured. Another larger fan is installed in a door or a window of the building (or the dwelling) containing the room (see Figure 4.3).

In order to get the leakage characteristics for a given element, the pressure in the room should be varied step by step from 10 Pa up to 60 or 70 Pa. The guarding pressure should be varied simultaneously to maintain a zero average pressure difference between the room and the building. This pressure difference actually varies between -1 and $+1$ Pa. A fit through several measurements provides the airflow rate corresponding to a zero pressure difference.

Several experiments are necessary to measure the other walls of the room by simply opening or closing various doors and windows. When a set of

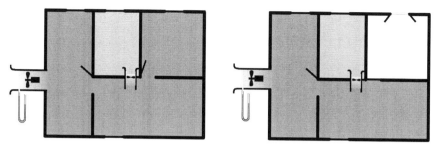

Figure 4.3 *Principle of the guarded zone technique applied to several walls of a room*

Note: At the left, only the external wall is measured, while at the right, the right partition wall is included in the measurement.

experiments is performed, enough equations can be written to compute the airflows through the various measured parts for each pressure step.

For that purpose, the pressure steps should be the same in each experiment, for example, 10, 30, 50 and 70 Pa. In order to get accurate pressure steps, the fans speeds are automatically controlled. Since even this control cannot be good enough to get accurate measurements (because of external and random influences such as temperature or wind fluctuations), the data should be automatically selected and recorded only when the following conditions are fulfilled:

- The pressure in the room is equal, within a predefined tolerance, to the predefined value of the pressure for each step.
- The pressure difference between the room and the guarding zone is smaller (in absolute value) than a predefined small value.

Finally, for each pressure step and each configuration, several values are measured and averaged to minimize the effect of random noise.

Determining the leakage coefficients

Density corrections

It is the airflow through the fan, q_m, which is measured and the airflow q through the leak that is needed to calculate the leakage coefficients. In pressurization experiments, the air blown by the fan comes from outside while the air leakage comes from inside. In depressurization experiments, the opposite is the case, but in both cases the temperature of the airflows may not be the same. During the measurements, the mass of air is conserved and:

$$\rho_m q_m = \rho q \qquad \text{hence} \qquad q = \frac{\rho_m q_m}{\rho} \qquad (4.4)$$

where ρ_m and ρ are the densities of the air going respectively through the fan and through the leaks. Since the density of the air is inversely proportional to its absolute temperature, and as long as the pressure differential remains small with respect to the atmospheric pressure:

$$q = q_m \frac{T}{T_m} \qquad (4.5)$$

where T and T_m are the absolute temperatures of the air going respectively through the measured elements and through the fan or the airflow-measuring device. This assumes that the variations of air moisture do not significantly change the density. Before any further analysis, Equation 4.5 should be used to correct the measured flows for density if the indoor–outdoor temperature difference is larger than a few degrees. Note that a difference of 10°C will induce a bias of 3 per cent in the airflow rate if this correction is not performed.

Two measurement points

If measurements are performed at two pressures only, for example at the lowest accurately measurable pressure differential and at the maximum acceptable one, results of measurements are Δp_1, q_1 and Δp_2, q_2. The coefficients of the power law are then:

$$n = \frac{\log q_1 - \log q_2}{\log \Delta p_1 - \log \Delta p_2} \qquad \text{and} \qquad C = \frac{q_1}{\Delta p_1^n} = \frac{q_2}{\Delta p_2^n} \qquad (4.6)$$

The coefficients of the quadratic law are:

$$a = \frac{\Delta p_1 q_2 - \Delta p_2 q_1}{q_1 q_2 (q_2 - q_1)} \qquad \text{and} \qquad b = \frac{\Delta p_1 q_2^2 - \Delta p_2 q_1^2}{q_1 q_2 (q_2 - q_1)} \qquad (4.7)$$

More than two measurement points

More than two measurements may be useful for testing the fitness of the model and to increase the accuracy of results. In this case, the least square fit method can be applied to get the coefficients of the power law. For this, Equation 4.1 can be linearized by taking the logarithm of both sides:

$$\log Q = \log C + n \log \Delta p \qquad (4.8)$$

This expresses a linear relationship between $\log Q$ and $\log \Delta p$ (see Figure 4.4):

$$y = a + nx \qquad (4.9)$$

with:

$$y = \log Q$$
$$a = \log C \qquad (4.10)$$
$$x = \log(\Delta p)$$

An appropriate fitting technique (see Chapter 7, 'Identification methods') can be used to identify the parameters, a and b, and the corresponding confidence

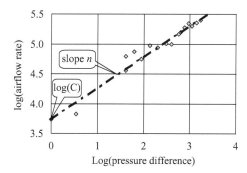

Figure 4.4 *Logarithmic plot of airflow rates and pressure differences*
Note: The slope of the best-fit line is an estimate of *n* and its ordinate at origin is an estimate of log(C).

intervals. If the coefficients a and b are known, the airflow coefficient C and the exponent n are calculated using:

$$C = \exp(a) \qquad \text{and} \qquad n = b \tag{4.11}$$

The Etheridge model in Equation 4.2 can be rewritten, dividing by the airflow rate q:

$$\frac{\Delta p}{q} = a + bq \tag{4.12}$$

that is again a linear model:

$$y = a + bx \tag{4.13}$$

with

$$y = \frac{\Delta p}{q} \qquad \text{and} \qquad x = q \tag{4.14}$$

In this case, the linear fit directly provides the coefficients a and b of the Etheridge model.

The measurement points can also be interpreted using the inverse problem theory (Tarantola, 1987), taking into account *a priori* knowledge such that the exponent n is between 0.5 and 1. Fürbringer *et al.* (1994) propose such a method, which has the advantage of providing a clear view of the error margins of the coefficients.

Corrections for standard conditions

Coefficients obtained from measurements that are performed under different atmospheric conditions should be corrected to reduce them to standard conditions, for example 20°C and 101,300 Pa.

Using the subscript o for these standard conditions and no subscript for the measurement conditions, then:

$$C_o = C \left(\frac{\mu}{\mu_o} \right)^{(2n-1)} \left(\frac{\rho}{\rho_o} \right)^{(1-n)} \tag{4.15}$$

where μ is the viscosity $(kg\,s^{-1}\,m^{-1})$, ρ the density of air and n the power law exponent. The variation of the air density is:

$$\frac{\rho}{\rho_o} = \frac{p T_o}{p_o T} \tag{4.16}$$

and the variation of the viscosity is given by the following approximation as a function of the absolute temperature, T:

$$\mu = \frac{1.458 \cdot 10^{-6} \sqrt{T}}{1 + \dfrac{110.4}{T}} \tag{4.17}$$

hence

$$\frac{\mu}{\mu_o} = \sqrt{\frac{T}{T_o}} \, \frac{1 + \dfrac{110.4}{T_o}}{1 + \dfrac{110.4}{T}} \cong \frac{17.1 + 0.047\theta}{17.1 + 0.047\theta_o} \tag{4.18}$$

where θ is the temperature in degrees Celsius. The approximation given in the second part of Equation 4.18 can be used between $-10°C$ and $40°C$.

Since the correction is small and if the temperatures and pressures are known with a reasonable accuracy, the additional errors introduced by this correction are negligible.

Ways of expressing the airtightness

For practical reasons, permeability is often characterized by one figure only. Some information is of course lost when one figure is used to represent the permeability instead of two. The following ways are commonly used for this issue.

Airflow rate at conventional pressure

The airflow rate at a given, conventional pressure, is calculated from Equations 4.1 or 4.3 depending on which parameters are available. The conventional pressure is usually 1, 4, 10 or 50 Pa, depending on the standard used or on the local uses.

50 Pa corresponds to a pressure differential commonly used for measurements and therefore at a pressure range for which the leakage rate is measured accurately. It does not, however, correspond to a typical pressure differential across building envelopes, which is closer to 4 Pa. Airflow rate at 1 Pa is the coefficient C in Equation 4.1. 10 Pa is a compromise between accuracy obtained at high pressures and actual, lower pressures.

Virtual air change rate

By dividing the airflow rate at conventional pressure by the internal volume of the tested enclosure gives a virtual leakage air change rate at that pressure. For this figure, 50 Pa is the most used pressure difference, and the figure is then noted $n_{50,}$ in $[\text{h}^{-1}]$. This value is less than $1\,\text{h}^{-1}$ in airtight buildings but, depending on the climate and building habits, buildings may have figures larger than $10\,\text{h}^{-1}$. This figure does not indeed characterize the quality of the envelope, since it depends on the volume of the enclosure. It provides an indication of the importance of infiltration in relation to building ventilation.

Specific leakage rate

The airflow rate at conventional pressure divided by the area of the envelope of the tested enclosure provides a figure characterizing this envelope. For such application, the most common pressure differential is 4 Pa, and this parameter is then v_4, or specific leakage rate at 4 Pa. It is expressed in $\text{m}^3/(\text{h}\,\text{m}^2)$. It is also the average air velocity through the envelope. This figure is less than 1, even $0.5\,\text{m}^3/(\text{h}\,\text{m}^2)$ for airtight envelopes.

Equivalent leakage area

An equivalent leakage area, i.e. the area of a circular hole with sharp edges that would have the same airflow rate at a given pressure differential, is:

$$A_L = C\sqrt{\frac{\rho}{2}}\,\Delta p^{(n-1/2)} \tag{4.19}$$

The uncertainty of the leakage area resulting from uncertainties on coefficients C and n is:

$$\delta A_L = \sqrt{A_L^2\left[\left(\frac{\delta C}{C}\right)^2 + (\delta n \ln \Delta p)^2\right]} \tag{4.20}$$

Specific equivalent leakage area

The equivalent leakage area can be divided by the area of the envelope of the tested enclosure to provide a specific leakage area. At 4 Pa, this ratio, expressed in cm^2/m^2, is close to the specific leakage rate expressed in $\text{m}^3/(\text{h}\,\text{m}^2)$.

Airtightness of buildings

The main reason for conducting building airtightness measurements is to characterize the leakage of the building envelope in the absence of climatic or other variable parameters influencing the results. Therefore the building (or part of the building or a particular component) is pressurized or depressurized in order to create a pressure difference large enough to minimize influences from wind and temperature differences on the results. This pressure differential is built up and maintained by means of a fan, forcing airflow through the envelope or component to be evaluated. This amplified airflow can be put in evidence by both qualitative (visualization) as well as quantitative (measurement of the airflow for a given pressure difference) techniques in order to assess the leakage locations, areas and characteristics.

External fan

The technique involves replacing an external door with a panel containing a powerful, variable speed fan. Several commercial blower doors are now available. These can be adjusted to fit snugly into any domestic doorframe. Airflow through the fan creates an artificial, uniform static pressure within the building. Internal and external pressure taps are made and a manometer is used to measure the induced pressure differential across the building envelope. It has become common practice to test buildings up to a pressure difference of 50 Pa.

Some means must also be provided to enable the volumetric flow rate through the fan to be evaluated. The aim of this type of measurement is to relate the pressure differential across the envelope to the airflow rate required to produce it (see 'Determining the leakage coefficients', above).

The general configuration for a pressurization/depressurization test is shown in Figure 4.5. The measurement procedure will depend upon the purpose of the test and the exact equipment used.

The airflow required to produce a given pressure difference under pressurization (airflow in) will not necessarily be identical to the flow required to produce the same pressure differential under depressurization (airflow out). This difference is mainly due to the fact that certain building elements can act as flap valves. For example, some types of window will be forced into their frames under pressurization while the reverse will be true for evacuation. This implies that the actual leakage area of the building envelope will be a function of the type of test conducted. Hence, ideally, the fan and flow measuring mechanism must be reversible.

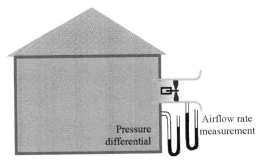

Figure 4.5 *Schematic of building airtightness test*

The overall airtightness of the structure and the size of the available fan govern the maximum volume of enclosure that may be pressurized. Even if large fans are available, in large leaky structures it may be possible to only achieve a limited range of pressure differentials. Several researchers have used trailer mounted fans with maximum flow capacities of about $25\,\mathrm{m}^3/\mathrm{s}$ to examine buildings with volumes as large as $50,000\,\mathrm{m}^3$.

Internal fan

Because of the size and cost of trailer-mounted equipment and the inherent difficulties of transportation and required manpower, other techniques have been developed for the examination of large buildings. One method is to create the required pressure differential using the building's existing air handling system. This technique relies on the building possessing a suitable mechanical ventilation system, which can be adjusted to meet the needs of the measurement. Essentially, the supply fans are operated while all return and extract fans are turned off and return dampers closed (or exhaust ducts sealed) so that the air supplied to the building can only leave through the leakage sites.

The analysis of measurement results proceeds along the same lines as that for small buildings, but because of the large building volume it may not be possible to achieve a pressure difference of 50 Pa.

Leakage visualization

Leakage can be visualized using infrared imaging, using a camera able to see far infrared radiation emitted by any surface. When reducing the internal pressure, outdoor air enters the building through leakage. Outdoor air, with a temperature that should differ from indoor air, changes the temperature of surfaces close to the leaks, thus making them visible, as in Figures 4.6 and 4.7 showing the connection between two walls and the roof of a wooden building. In this

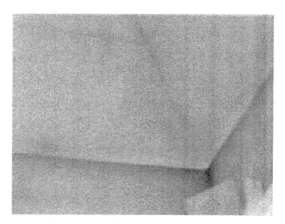

Figure 4.6 *Roof corner from inside*
Source: Roulet, 2004.

Figure 4.7 *Roof corner under depressurization*
Source: Roulet, 2004.

case, the airtightness is not good enough and cold air enters the inhabited space through cracks between wooden panels.

The stack effect method

This simple and easy-to-install method to estimate the air leakage distribution in tall buildings is based on the pressure distribution induced in buildings by the stack effect (Tamura and Wilson, 1966). Three parts can be estimated separately: the ground floor, the top floor and the remaining floors.

The basic idea is to pressurize the building with the stack effect, and to plan three different experiments where two airflows can be measured to get three independent equations for the three different leakages that will be estimated (Hakajiwa and Togari, 1990).

For this measurement method, the building should be tall and the temperature difference between indoors and outdoors should be large enough, in such a way that the pressure difference between inside and outside induced by the stack effect is larger than the pressure caused by the wind. Therefore, calm weather should be preferred and the mechanical ventilation system switched off. The pressure difference resulting from buoyancy is proportional to the product of the indoor–outdoor temperature difference and maybe the stack height. It reaches 30 Pa if the product of the height and the temperature difference is 700 Km.

The leakage of the building is divided into three parts:

- leakage through the ground level including the entrance door (suffix g);
- leakage through the top level including the roof (suffix t);
- leakage through the remaining floors (suffix r).

If the building has all its internal doors open as well as the staircase and the lift shaft, and if the temperature does not vary too much throughout the building, there is *a priori* only one neutral plane at the height z_0. The neutral plane is the generally horizontal plane in the building or part of it where the indoor–outdoor pressure differential is zero. Its height depends on the size and position of the ventilation and leakage openings. It is such that the airflows going in and out of the building are balanced.

The pressure difference, Δp, caused by the stack effect at any height, z, in a given building configuration is then:

$$\Delta p(z) = \int_{z_0}^{z} \Delta \rho(z) g \, dz \tag{4.21}$$

where $\Delta \rho(z)$ is the difference between the densities of indoor and outdoor air at height z, and $g = 9.81 \text{ m/s}^2$ is the acceleration due to gravity.

If the temperatures are homogeneous, Equation 4.21 gives:

$$\Delta p(z) = \Delta \rho g (z - z_0) \tag{4.22}$$

Using the law of perfect gases to express the air density, we get:

$$\Delta p(z) = \frac{Mp}{R} \left[\frac{1}{T_i} - \frac{1}{T_e} \right] g(z - z_0) \tag{4.23}$$

where:

T_i and T_e are the indoor and outdoor air absolute temperatures,
M is the average molar mass of the air, i.e. 0.029 kg/mole,
p is the atmospheric pressure,
R is the constant for perfect gases, i.e. 8.31396 J/mole · K.

The leakage of the building is represented by the usual power law:

$$Q = C \, \Delta p^n \tag{4.24}$$

Assuming that the exponent n is the same for every leak, there are three unknowns, the leakage coefficients, C_g, C_r and C_t. To estimate these coefficients, three measurements are performed, where the pressure differences, the temperatures at various heights in the building and some airflows are measured. A first relationship is given by the conservation of mass with a closed envelope. The two other equations are obtained by mass conservation with a large opening at the bottom and at the top of the building. In these cases, the airflows through these openings are measured. The relations are as follows:

1   All openings closed – in this case, the neutral plane is somewhere at mid-height of the building and, by conservation of the mass of air, we have:

$$\rho_g C_g \, \Delta p(z_g)^n = \rho_t C_t \, \Delta p(z_t)^n + \int_{z_{rb}}^{z_{rt}} d(\rho_r C_r \, \Delta p(z_r)^n) \tag{4.25}$$

where ρ_g, ρ_r and ρ_t are the densities of the air at the ground level, the remaining floors and at the top level in order to have the proper mass flow.

2   Entrance door open – the airflow through the open entrance door (or any other large opening on the ground level), Q_g, is measured, either by

measuring the air speed at several locations and integrating over the whole opening or using a tracer gas.

$$\rho_g Q_g + \rho_t C_t\,\Delta p(z_t)^n + \int_{z_{rb}}^{z_n} d(\rho_r C_r\,\Delta p(z_r)^n) = 0 \qquad (4.26)$$

3   Windows open at the top level – the airflow through these windows, Q_t, is measured. We have similarly:

$$\rho_g C_g\,\Delta p(z_g)^n + \rho_t Q_t + \int_{z_{rb}}^{z_{rt}} d(\rho_r C_r\,\Delta p(z_r)^n) = 0 \qquad (4.27)$$

The neutral plane is now at the top level.

Assuming that n is 0.6 or two-thirds, which are the most probable values, the system of three equations above can be solved to estimate C_g, a global C_r and C_t. If the temperatures are not uniform inside or outside, Equation 4.21 should be used instead of Equation 4.22. The system is then more complex but can still be solved. The most important condition to observe during the measurement is the absence of wind.

The main advantage of the method is that it does not require the use of sophisticated equipment. As a minimum, the required equipment is:

- wind velocity meter, 0–5 m/s, for measuring the airflow rate in the openings;
- differential manometers, 0–50 Pa;
- air temperature thermometers;
- length measuring device as long as the building is tall.

This equipment can be completed by more differential manometers and more thermometers, used to verify the linearity of the pressure distribution through the building.

Neutral height method

A simple variant of the stack effect method offers in many cases a good estimate of the leakage area and determines if specifications are met or exceeded (Van der Maas *et al.*, 1994). It is also based on the determination of the neutral height and the equipment necessary is only an airflow direction detector (small smoke generator such as a cigarette, incense stick or small flame) and a yardstick. With no wind, this method even allows the leakage characterization of a single storey building.

The measurement should be performed with the mechanical ventilation system switched off, preferably on a cold day without wind and when the building is heated. In these conditions, the airflow through the building results from buoyancy only. The method can also be used, *mutatis mutandis*, in cooled buildings in a warm climate.

The principle is to determine the position of the neutral height in an outside opening such as a door. The airflow direction detector is moved from bottom to

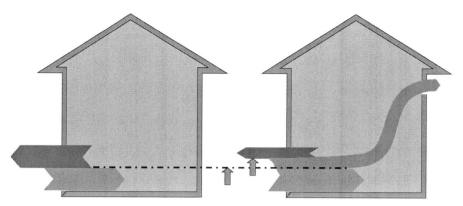

Figure 4.8 *Principle of the neutral height method for assessing leakage area*

Note: The left shows no leak (except the test opening), while on the right, the leakage is above the test opening.

Source: Roulet, 2004.

top of the opening to observe the flow direction. The neutral level is located between the ingoing and outgoing flow directions. Sensitivity can be increased or decreased by reducing or increasing the width of the opening.

In an airtight building, the neutral height will be located at about mid-height of the measuring opening (see Figure 4.8, left). Cold air enters the building through the lower half of the opening, and warm air leaves the building through the upper part.

Using the Bernoulli equation and mass conservation, it can be shown that the mass flow rate through the upper or lower half of the opening is:

$$Q_A = \frac{1}{3} C_d \rho_o A \sqrt{\frac{gH\,\Delta T}{T_0}} \tag{4.28}$$

where (Van der Maas *et al.*, 1994):

C_d is the discharge coefficient of the opening (for example, 0.6),

ρ_o is the density of air outdoors (assumed to be the cold zone at temperature T_o),

A is the area of the opening of with W and height H,

g is the gravitational acceleration (about 9.81 m/s),

ΔT is the indoor–outdoor temperature difference.

If some air enters or leaves the enclosure through another opening or through leaks, part of the incoming air will leave or enter the building through it and will not, therefore, pass through the upper or lower part of the test opening. The neutral height rises or goes down to balance the two airflows (see Figure 4.8, right). Using mass conservation, the net mass flow rate is given, assuming that $T_i > T_o$, by:

$$Q = Q_A \left(\frac{T_o}{T_i}\right)^{1/2} \left(1 - \frac{z_n}{H}\right)^{3/2} - \left(\frac{z_n}{H}\right)^{3/2} \cong Q_A(1-a)^{3/2} - a^{3/2} \qquad (4.29)$$

with $a = \dfrac{z_n}{H}$

where z_n is the height of the neutral level.

The opening area between this neutral level and the mid-height of the opening is close to the equivalent leakage area. If the leakage is small, the sensitivity can be increased by reducing the width of the opening, for example by partly closing the door.

A neutral height below the mid-height of the test opening means that most of the leakage area is below the opening. If the other leakages or openings are larger than the test opening, the neutral height will not be found within the test opening, even when this is wide open. In this case, a walk through the building is necessary to identify and, if possible, to close or seal these large openings.

The equivalent area measured this way is the difference between the equivalent areas of the openings or leakage areas located above and below the opening. Therefore, it is useful to make this measurement at two test openings located at the bottom and the top of the building.

Measurement of airtightness of a duct or network

To ensure that fresh air reaches the ventilated space, thus ensuring acceptable air quality, and to avoid energy waste when the air is either heated or cooled, the duct network should be airtight. Significant energy may be wasted, for example, where leaky ductwork passes through an unheated space such as an attic, basement or crawl space. As an example, it was found that ductwork is the most significant source of leakage in western US houses, together with fireplaces (Dickerhoff *et al.*, 1982). Modera (1989) confirms these findings, but some houses were nevertheless found to be acceptable.

Several techniques allow for the checking of the airtightness of air ductwork. In Finland, the airtightness of the ventilation system has to be checked when commissioning the system (NBCF, 1987), but in most countries, measurements are seldom carried out. ASTM (2003) provides guidance to perform such tests. Some measurement methods are described below.

Pressurization method

The principle of this method is a combination of those principles described in Chapter 2, 'Measurement for airflow rate in a duct' and Chapter 4, 'The fan pressurization method', above. All intakes, supply terminals, exhaust and extract terminals connected to the system should be carefully sealed, for

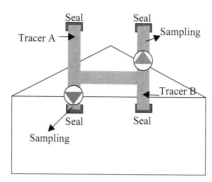

Figure 4.9 *Location of tracer injection and sampling tubes for the measurement of leakage airflow rates in a ventilation system*
Source: Roulet and Vandaele, 1991.

example, using plastic sheeting and adhesive tape. Inflated balloons are also well suited to seal circular ducts.

Tracer gas injection and air sampling tubes are installed at appropriate points in the main supply or exhaust ducts, as shown in Figure 4.9, to quantify any residual flow rate resulting from leakage.

The system fans (or a fan added at one register if required) are used to pressurize the supply side and depressurize the exhaust side of the network. Since all inlet and exhaust grilles or ducts are sealed, the flow, q_L, through the fan(s) results from leakage, and is measured as described in Chapter 2, 'Measurement of the airflow rate in a duct', together with the pressure difference, Δp, between the inside and outside of the ducts. The flow rate is the sum of all leaks downstream of the measurement point in pressurized ducts, and upstream for depressurized ducts.

A series of measurements is made at different fan speeds, and the coefficients of Equations 4.1 or 4.2 are determined, and the relationship is subsequently used to calculate the leakage rate at the service pressure difference.

Flow rate difference method

If a duct is very leaky, the leakage can be obtained by measuring the difference between the flow rates at two locations along the flow. Since additional pressure drop should be avoided, tracers are recommended to measure the airflow rates. One tracer should be injected at a point upstream of the first location, and a second tracer injected at the first location. The concentrations of each tracer are measured after the second, downstream location, at a distance where a good mixing is achieved (see Chapter 2, 'Sampling points for concentration measurements'). If steady flows can be assumed, two sequential measurements using a single tracer at each point may be used instead.

For depressurized ducts, only one tracer is necessary since it is diluted by the air entering the duct through leakage. The tracer gas is injected at the upstream end of the duct and its concentration is measured at both ends to give the flow rate at each.

The leakage of the whole supply or exhaust network may be determined by measuring the difference between the airflow rate in the main duct (close to the fan) and the sum of all the flow rates at the individual inlet or extract terminals. For this purpose, the main airflow rate through the fan may be measured with a tracer, and the flow rates at the terminals may be determined by one of the methods described in Chapter 2, 'Airflow measurements in air handling units', the most appropriate being the compensated flowmeter (see Chapter 2, 'Compensated flowmeter').

With this method, the leakage flow rate is the difference of two large numbers. Therefore, it is not the best one for tight or only slightly leaky ducts. Because of its ease of use, it can nevertheless be used for diagnosis purposes, to detect if the ductwork is very leaky or not.

Differential building pressurization

The methods described above measure the leakage of the air duct system. From the point of view of energy saving, however, it may be useful to measure the air leakage to or from outside only, and not leakages between the system and the interior of the building.

For this purpose, the duct system is assumed to be a part of the envelope and the duct leakage is obtained by difference. In a first test, the closed building is pressurized after sealing the outdoor air intake and exhaust of the building, with all the registers and returns open (see Figure 4.10, left). In a second test, all registers and returns are sealed (see Figure 4.10, right). The difference in airflow rate between the two tests, for each pressure difference, results from duct leakage to the outside.

The major advantage of this method is that it needs little equipment, in addition to that required for envelope leakage measurements. However, it is prone to inaccuracy since the duct leakage is again obtained as the difference between the measurements of two large airflow rates.

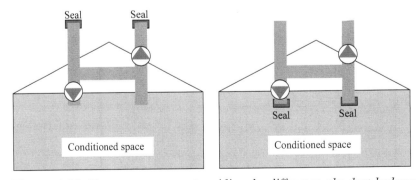

Figure 4.10 *Two measurements providing, by difference, the duct leakage to outside of the conditioned space*

5

Measurements and Measures Related to Energy Efficiency in Ventilation

Energy in buildings

Energy uses and indoor environment quality

Energy is used in buildings for many purposes such as:

- heating and cooling;
- drying and humidifying;
- ventilation (moving the air);
- hot water supply;
- lighting;
- building systems such as lifts, escalators, communication networks;
- cooking, washing, leisure, producing goods and services.

According to the Rio Declaration, sustainable buildings should take account of environmental, economical and social factors. This includes, among others, low energy use, good indoor environment quality and health. The three factors have equal importance: a building cannot be good if it fails in one of them. Ventilation plays a large role in these issues by ensuring a good indoor air quality. In mechanically ventilated buildings, ventilation uses energy to move the air and, in many cases, to condition it.

In some case, especially when appropriate studies are not performed, there may be a conflict between strategies to reduce energy use and to improve indoor environment quality. However, studies and existing high performance buildings show that it is possible to realize healthy, comfortable and energy efficient buildings. Basic recommendations to reach these objectives are (Roulet, 2004):

- Prefer passive methods to active ones wherever possible.
- Think about user comfort, needs and behaviour.
- Adapt the building and its systems to its environment.

Passive and active ways to get high quality buildings.

Passive ways are architectural and constructive measures that naturally provide a better indoor environment quality without or with much less energy use. Examples are:

- improving winter thermal comfort with thermal insulation, passive solar gains, thermal inertia and controlled natural ventilation;[1]
- improving summer thermal comfort with thermal insulation, solar protection, thermal inertia and appropriate natural ventilation;
- ensuring indoor air quality by using low-emitting materials and controlled natural ventilation;
- providing controlled daylighting;
- protecting from outdoor noise with acoustical insulation adjusting the reverberation time for a comfortable indoor acoustics.

Passive means are often cheap, well accepted by the occupants, use very little or no energy, and are much less susceptible to break down than active means. However, they often depend on meteorological conditions and therefore cannot always fulfil their objectives. They should be adapted to the location and therefore need creativity and additional studies from the architect, and a design error may have dramatic consequences.

Active (or technological) ways improve the indoor environment quality by mechanical actions, using energy to complement the passive ways or even to compensate for low building performance. Examples are:

- heating boilers and radiators for winter comfort;
- artificial cooling by air conditioning or radiant panels for summer comfort;
- mechanical ventilation;
- artificial lighting;
- actively diffusing background music or noise to cover the ambient noise.

Active ways, when appropriately designed, built and maintained, are perfectly adapted to needs. Flexible and relatively independent of meteorological conditions, they allow for the correction of architectural errors. However, the required technology is often expensive, uses a lot of energy and may break down. Furthermore, active means require a higher maintenance input.

Passive ways are preferred, but cannot always fulfil the comfort objectives. Therefore, the appropriate strategy is to use them as much as reasonably possible and to compensate for their insufficiencies with active systems, which will then be of lesser importance. This strategy often allows more freedom in choosing the type and location of active systems.

The passive way of ensuring indoor air quality is of course natural ventilation, but also reduction of pollutant emission indoors by an appropriate choice of building materials and furniture. The corresponding active way is mechanical ventilation wherever necessary, including heating, cooling, humidification or dehumidification. An appropriate design of the ventilation systems, a careful commissioning of new systems and conscientious maintenance guarantee good indoor air quality at a lower energy cost. Once again, measurements may help in commissioning and diagnosing failures.

Energy in air handling units

Energy for heating and cooling buildings

Heating and cooling aim to keep a quasi steady and comfortable temperature indoors, despite variations of the outdoor temperature, and taking into account the solar radiation heating the building fabric, mainly through windows, and internal heat gains from occupants and their activities. The amount of energy needed for this depends on the following parameters:

- the climate, which is the imposed boundary condition;
- the opaque parts of the building envelope, the function of which is to protect the indoor environment against the weather. Reinforced thermal insulation and good airtightness are essential for this purpose;
- the transparent parts of this envelope, ensuring daylighting and view, but also allowing the solar radiation to heat the indoor environment. This passive solar heating is welcome in the cold season, but induces overheating in the warm season. Therefore, transparent parts of the envelope should be equipped with mobile and efficient solar protection to control the passive solar heating and daylighting;
- the thermal inertia (thermal mass) of the building fabric, which naturally stabilizes the indoor temperature;
- the internal gains resulting from occupants' activities, which contribute to heating in the cold season but add to the cooling load in hot season.

It can easily be seen that heating and cooling energy needs depend mainly on the building design and its location. These energy needs may be satisfied by different systems, including air conditioning. The systems should be energy efficient, i.e. satisfy the needs at a minimum energy cost. Among the numerous heating and cooling systems, we only consider here using air as a medium. It should be mentioned that, because of its low density, air is a poor heat carrier: one litre of water carries, in practice, 7–15 times more heat than one cubic metre of air![2]

Energy for air conditioning

Buildings are primarily ventilated for the purpose of removing the pollutants generated within them. The air leaving the buildings has the characteristics (temperature, humidity, chemical composition) of the indoor air. It is replaced at the same mass airflow rate by air coming from outdoors, which also has its own characteristics. Air conditioning is giving or taking heat and water vapour to or from the outdoor air entering the building to obtain the required indoor air temperature and humidity. This needs energy.

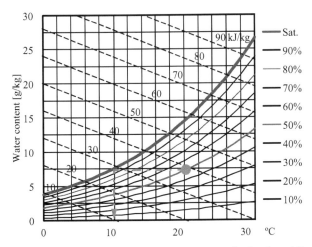

Figure 5.1 *Psychrometric chart with constant relative humidity curves and constant enthalpy lines*

Note: It is shown that air at 20°C and 50 per cent relative humidity contains about 7.5 g of water vapour per kilogram. Its enthalpy is 39 kJ/kg and its dew point is close to 10°C.

Figure 5.1, a psychrometric chart for air, shows several characteristics of humid air. The curves show the water content of air as a function of its temperature for various relative humidities. The water content cannot be greater than that shown by the saturation curve. Air with a relative humidity $0 < \varphi < 1$ has a water content φ times that of the saturated air.

Energy is needed for heating or cooling the air as well as for evaporating water in it or condensing water for drying it. Taking as a reference dry air at 0°C, the specific enthalpy or energy needed to heat and humidify 1 kg of air to reach the temperature θ and humidity ratio x is:

$$h = c_{da}\theta + (L + c_w\theta)x \tag{5.1}$$

where:

c_{da} is the specific heat capacity of dry air, about 1006 J/(kgK);
c_w is the specific heat capacity of water vapour, about 1805 J/(kgK);
L is the latent heat of evaporation, i.e. the heat required to evaporate 1 kg of water, about 2,501,000 J/kg;
x is the humidity ratio, i.e. the mass of water vapour per kilogram of dry air.

This humidity ratio, x, is related to the water content, ω, which is the mass concentration of water vapour in moist air of Figure 5.1, by:

$$x = \frac{\omega}{1 - \omega} \quad \text{and} \quad \omega = \frac{x}{1 + x} \tag{5.2}$$

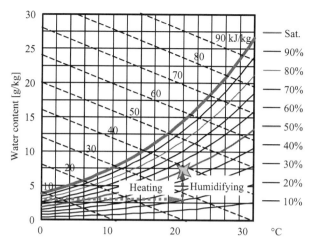

Figure 5.2 *Paths in the psychrometric chart for heating and humidifying outdoor air in winter to reach 20°C and 50 per cent relative humidity*

Figure 5.2 illustrates the paths of temperature and water content of air for heating and humidifying outdoor air in winter, at 0°C and 80 per cent relative humidity, in order to get 50 per cent relative humidity at 20°C.

A part of the energy required for heating and humidifying the air in winter is brought from free sources such as solar radiation or metabolic activity of occupants. All the electricity used for lighting and other appliances that are not part of the heating system end up as heat, in most cases released into indoor air. Plants and occupants, as well as activities such as cooking and drying laundry add water vapour to indoor air. When this is not enough to reach a comfortable indoor climate, the complement is provided by a heating system. In this case, about 0.34 Wh is needed to heat or cool 1 m³ of air by 1°C, as long as the air is humidified by 'free sources'. This value is therefore used in models calculating the energy for heating buildings. If a humidifier is used, it will take 2.5 kJ (about 0.7 Wh) per gram of water vapour generated. This heat is taken in the indoor environment if the humidifier does not generate water vapour but water droplets (spray humidifiers).

The paths in the psychrometric chart for cooling and drying summer outdoor air from 30°C and 70 per cent relative humidity down to 20°C and 50 per cent relative humidity are shown in Figure 5.3. Note that, for drying the air, it should first be cooled down at the dew point temperature corresponding to the required specific humidity, and then reheated to the required indoor temperature. It should also be noticed that it is impossible to cool the air below its dew point without drying it.

Hot and humid outdoor air cools down and eventually dries on contact with cold surfaces, on which excess water vapour condenses. If these surfaces are not cooled down, such as the building fabric or furniture, their temperature rises and cooling stops after a while. However, the air temperature rises more

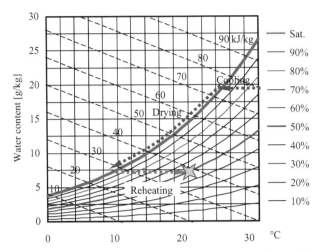

Figure 5.3 *Paths in the psychrometric chart for heating outdoor air in winter or cooling it in summer to reach 20°C and 50 per cent relative humidity*

slowly if the air is in contact with massive structures that were cooled down before, for example, by strong airing during the cool night.

Mechanical cooling is needed to keep the surfaces in contact with the air cold, and to get continuous air drying and cooling. Warm, humid air is first cooled down when passing through a refrigerated heat exchanger (horizontal 'cooling' line in Figure 5.3) until it reaches its dew point. Then it is dried by losing the water that condenses on the heat exchanger ('drying' curve) until it reaches the required specific humidity, at a new lower dew point. It should then be reheated to the required temperature.

Numerical values for this process are given in Table 5.1. The largest change in enthalpy is when drying, since 2500 J should be withdrawn from the heat exchanger to condense each gram of water.

The energy required to reheat the dry, cold air can be provided by various means:

Table 5.1 *Humidity ratio and specific enthalpy of warm, humid air cooled down and dried as shown in Figure 5.3*

Process	Temperature θ (°C)	Relative humidity φ (%)	Humidity ratio x (g/kg)	Specific enthalpy h (J/(kg · K))	Enthalpy increase Δh (J/(kg · K))
Cooling	30.0	70	18.8	78,756	−6941
Drying	23.9	100	18.8	71,815	−44,244
Heating	9.3	100	7.3	27,571	11,153
	20.0	50	7.3	38,724	

- From the indoor environment, heat loads and solar gains. This way, common in tropical climates, saves the investment of the heating system, and heating energy is free. It has, however, the disadvantage of blowing cold air into the occupied spaces, often leading to draughts. In such systems, recirculation is often very large and temperature control is obtained by varying the supply airflow rate.
- Heat provided to a warm heat exchanger by the heat pump used to cool down the chilled water. This heat pump provides cooling water at temperatures higher than indoor temperature. This water or a part of it can be circulated into the warm heat exchanger without any running cost. The investment is limited to pipes connecting the chiller condenser to the warm heat exchanger and to a control valve.
- Heat provided to a warm heat exchanger by a separate heating system. This is expensive both in investment and running costs and should not be used.

Measurement of energy for heating, cooling, humidifying or dehumidifying air

The amount of energy needed to increase the temperature and humidity of a known volume of air depends only on the start and final values of temperature and humidity ratios. Using Equation 5.1:

$$Q = h\rho V = \rho V c_{da}\theta + (L + c_w\theta)x \qquad (5.3)$$

Therefore, measuring the airflow rate (according to Chapter 2, 'Measurement of airflow in a duct') through the heating coils and humidifier (if any), as well as air temperature and moisture upwind and downwind of these elements, allows for the calculation of the heating and humidifying power.

This is not that simple for cooling and dehumidifying. The measurement of airflow rate is the same, but air temperature and humidity should be measured before and after each of the processes mentioned in Table 5.1:

- Cooling and dehumidification – measurements in outdoor air and after the cooling coils provide the power taken from the chilled water. This power may also be obtained by measuring the chilled water flow rate in the cooling coil and its temperature increase.
- Reheating – measurements before and after the heating coil give the power provided by the reheating system. This power can also be calculated from measurements of the heating water flow rate in the heating coil and its temperature decrease.

Heat exchangers

The purpose of heat exchangers is to transfer heat from water to air (heating coils) or vice versa (cooling coils). This heat should be transferred in the most efficient way possible, without transferring contaminants. The diagnosis should characterize the performance of the exchanger.

To improve energy efficiency, mechanical ventilation systems are often equipped with heat recovery for recovering the heat contained in exhaust air. This heat is in most cases given back to supply air. Such heat recovery exchangers are efficient during both cold and hot seasons, saving heating and cooling energy. Some of these heat exchangers also transfer humidity, thus decreasing the energy used to humidify or dehumidify the air.

As shown in Figures 0.3, 0.4 and 0.5, air handling units may have parasitic shortcuts and leakages. Such leakages have often been observed in buildings (Presser and Becker, 1988; Hanlo, 1991; Fischer and Heidt, 1997; Roulet *et al.*, 1999). They can dramatically decrease the efficiency of ventilation and heat recovery (Roulet *et al.*, 2001). Moreover, leakage in a building's envelope allows indoor air to escape outdoors without passing through the heat recovery system. In addition, these units use electrical energy for fans, which may, in some cases, exceed the saved heat. The influence of these various phenomena on the real energy saving is addressed in this chapter.

Types of heat exchangers

Water-to-air heat exchangers are in most cases made out of finned tubes in which the water circulates. The fins increase the exchange area between the exchanger surface and the air.

The heat exchangers most commonly used for heat recovery are plate heat exchangers, rotating heat exchangers and heat pipes. Most common air-to-air exchangers are plate heat exchangers, in which the exhaust air is blown in several channels limited by plates made of glass, metal or plastic (see Figure 5.4). The other side of these plates is in contact with inlet air, so that heat

Figure 5.4 *Close view of a flat plate heat exchanger*

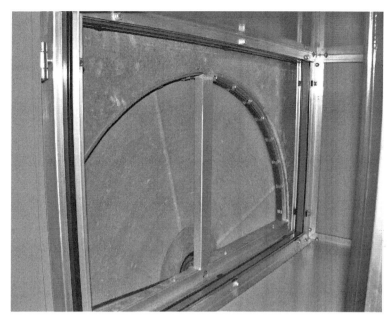

Figure 5.5 *Top half of a rotating heat exchanger*

can be transferred from the warm side to the other. The heat recovery efficiency of these exchangers ranges from 60 to 80 per cent, depending on the type and size. A variant of this exchanger is the heat pipe exchanger, in which heat pipes are used to transport heat from warm to cold air. The air leakage between both sides of such heat exchangers should be zero.

Rotating heat exchangers

Rotating heat exchangers are used in larger systems (see Figure 5.5). A disc with a porous structure (honeycomb, corrugated metallic foils) allowing the air to flow easily through it, is placed so as to have half of its area in the exhaust duct, and the other half in the supply duct. This disc rotates slowly and is heated in the warmer air, where air moisture may also condense on the surface of the porous structure. It is then cooled in colder, dryer air, also evaporating here the condensed water. This way, sensible and latent heat contained in warm air is given to cold air, and the heat recovery efficiency may reach 90 per cent. A gasket and a purging sector limit contamination from exhaust air to fresh air, without eliminating it completely (see Chapter 6, 'Contaminant transport in rotating heat exchangers').

A small leakage can be accepted in a rotating heat exchanger, resulting in a recirculation rate of less than 4 per cent. Reduced leakage is achieved by carefully installing the rotating heat exchanger, and by balancing the air pressure between both sides of the exchanger. To achieve this, supply and exhaust fans should not be on the same side of the heat exchanger (see Figure 5.6).

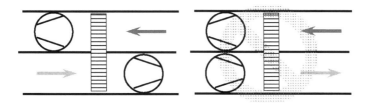

Figure 5.6 *Relative position of fans and rotating heat exchangers*

Placing both fans on the same side results in a large pressure differential through the rotating heat exchanger, thus increasing leaks. A parasitic recirculation rate as large as 40 per cent was measured by the author in such a unit!

Most rotating heat exchangers are equipped with a purging chamber, located between inlet and exhaust air ducts, on the warm side of the wheel (see Chapter 6, 'Contaminant transport in rotating heat exchangers').

Glycol heat exchanger

When exhaust and inlet ducts are not side by side, heat can be transported by a hydraulic circuit with two heat exchangers. The fluid (generally a glycol–water mix) is heated by the air–liquid heat exchanger located in one of the ducts, then pumped to the other exchanger to give heat to the cold air.

Heat pump

In exhaust only systems, the recovered heat cannot be given to outdoor air, but to the hydraulic heating system or to a hot water boiler. For this, the temperature of the hot side of the recovery system is increased using a heat pump, whose cold source is the exhaust air.

Heat exchange efficiency

The efficiency of heat recovery exchangers has two aspects: the energy (or enthalpy) efficiency and the temperature efficiency.

The first is the ratio of the enthalpy flow delivered to the supply air by the enthalpy flow in exhaust air:

$$\eta_E = \frac{H_{\text{downwind, supply}} - H_{\text{upwind, supply}}}{H_{\text{upwind, exhaust}} - H_{\text{outdoor air}}} \tag{5.4}$$

If supplied air upwind of the heat exchanger (inlet air) has the same characteristics as that of the outdoor air, $H_{\text{outdoor air}}$ may be replaced by $H_{\text{upwind, supply}}$. The enthalpy of air is determined by its temperature and moisture content (Equation 5.3). Therefore, measurement of temperature and moisture content of air upwind and downwind of both sides of the heat exchanger allows the determination of the enthalpy efficiency of the heat exchanger itself.

The enthalpy flow, H, is the product of mass airflow rate and specific enthalpy, h:

$$H = \rho Q h \tag{5.5}$$

where ρ is the density of air.

At ambient temperature, a numerical expression of Equation 5.3 for air is:

$$h = 1004.5\theta + x(2,500,000 + 1858.4\theta) \tag{5.6}$$

where:

θ is the air temperature,
x is the humidity ratio, that is the mass of water vapour per kg dry air.

The humidity ratio can be calculated from water vapour partial pressure, p, and atmospheric pressure, p_a:

$$x = \frac{0.62198p}{p_a - p} \tag{5.7}$$

The water vapour partial pressure is calculated from relative humidity, φ by:

$$p = \varphi p_s \tag{5.8}$$

where p_s is the water vapour pressure at saturation, which depends on the temperature:

$$\text{if } \theta < 0 \qquad p_s = 610.5 \exp\left(\frac{22.5 \cdot \theta}{273 + \theta}\right) \tag{5.9}$$

$$\text{if } \theta > 0 \qquad p_s = 610.5 \exp\left(\frac{17.27 \cdot \theta}{237.3 + \theta}\right) \tag{5.10}$$

The humidity ratio can also be derived from mass concentration of water, C_w, or volume concentration, c_w:

$$x = \frac{C_w}{1 - C_w} = \frac{\rho c_w}{1 - \rho c_w} \tag{5.11}$$

Also interesting, and much simpler to assess, is the efficiency or effectiveness, or temperature efficiency of the heat exchanger, which reveals how well a heat exchanger performs. This efficiency is simply calculated from temperature measurements in both circuits of the heat exchanger:

$$\text{Hot side:} \qquad \varepsilon_{\theta,h} = \frac{\theta_{\text{hot, in}} - \theta_{\text{hot, out}}}{\theta_{\text{hot, in}} - \theta_{\text{cold, in}}} \tag{5.12}$$

$$\text{Cold side:} \qquad \varepsilon_{\theta,c} = \frac{\theta_{\text{cold, out}} - \theta_{\text{cold, in}}}{\theta_{\text{hot, in}} - \theta_{\text{cold, in}}} \tag{5.13}$$

When the mass flows multiplied by the specific heats are equal on both sides the efficiency will also be equal.

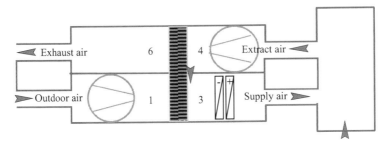

Figure 5.7 *Schematics of an air handling unit, showing location of pressure taps for pressure differential measurements*

Leakage through heat exchangers

Some heat exchangers let some air leak between both air channels. This is in most cases not expected, since there are very few air handling units equipped with both recirculation and a heat exchanger. In addition, some air is entrained by the rotation of the wheel in rotating heat exchangers. The amount of air transferred this way can be measured with tracer gases (see Chapter 2, 'Airflow measurements at ventilation grilles'), and the leakage flow rate is one of the results of the measurement of airflow rates in the air handling unit.

As mentioned in Chapter 2, the global recirculation rate can easily be checked by measuring the concentration of a tracer injected in the ventilated space, such as the carbon dioxide exhaled by occupants. Assuming that there is no inverse recirculation and no leaks in the air handling unit, the global recirculation rate is:

$$R = \frac{C_{\text{supply}} - C_{\text{outdoor}}}{C_{\text{exhaust}} - C_{\text{outdoor}}} \tag{5.14}$$

If no recirculation is expected, but a significant recirculation rate is observed, it may be the result of leakage through the heat exchanger. If more information is required, in particular to check whether it is the exchanger or another part of the air handling unit that leaks, additional measurements could be performed, as described in Chapter 2, 'Airflow rate measurements in air handing units'.

Pressure differential measurements are useful to explain leakage. In addition, these are easier to perform than leakage measurements and can readily bring information for a diagnosis. Pressure differentials should be measured between the following locations (see Figure 5.7):

- Between one and six on one hand, and three and four on the other hand. These pressure differentials drive the leakage direction. They should be zero or slightly positive, so that a possible leakage flow goes from supply to exhaust, and not the contrary.
- Between one and three on one hand, and four and six on the other hand. These pressure differentials increase with clogging. Compare them with the nominal pressure differential given by the factory for the actual airflow

rates. If these pressure differentials are significantly larger than the nominal values, the wheel should be cleaned.

Indication on how to measure pressure differentials is given in 'Measurement of pressure differences', below.

A word of caution: there should be no fan between the pressure taps used to measure the pressure differentials!

Effect of leakages and shortcuts on heat recovery

Definitions of global heat recovery efficiency

Building leakage and shortcuts within the ventilation system may significantly reduce the effectiveness of the heat recovery as shown below (Roulet *et al.*, 2001).

Consider the airflows in the ventilation unit schematically presented in Figure 5.8. Outdoor air enters the inlet grille and is blown through the heat recovery system, where it is either heated or cooled. When heat recovery is not needed, for example to bring free cooling during the night, plate heat exchangers are bypassed or the wheel is stopped in rotating heat exchangers.

Then, after additional heating or cooling when required, the outdoor air enters the supply duct to be distributed into the ventilated space. As the envelope is not perfectly airtight, the supply air is mixed with infiltration air in the ventilated space. A part of the air may also be lost by exfiltration. The extract air flows through the other part of the heat recovery system where it is either cooled (if inlet air should be warmed up) or heated (if fresh outdoor air should be precooled). The air is then blown to the outside through the exhaust duct to the atmosphere.

If the exhaust and inlet grilles are not well located, it is possible that a part of this exhaust air re-enters the inlet grille, resulting in an external recirculation rate. Leakage through the heat recovery system may also result in an internal recirculation rate, from inlet to exhaust, or from extract to supply.

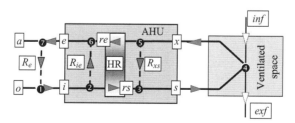

Figure 5.8 *The simplified network representing the air handling unit and ducts*

Note: *o*: outdoor air; *i*: inlet grille; *s*: supply air; *x*: extract air; *e*: exhaust air; *a*: atmosphere; HR: heat recovering exchanger; R_e: external recirculation; R_{ie}: inlet to extract recirculation; R_{xs}: extract to supply recirculation; *inf*: infiltration; *exf*: exfiltration. Arrows represent considered airflow rates.

Source: Roulet *et al.*, 2001.

In simplified methods to calculate heating (or cooling) demand of buildings, ventilation heat loss, Φ_V, is often calculated by (CEN, 1999, 2007):

$$\Phi_V = \dot{m}(h_x - h_o)(1 - \eta_G) \tag{5.15}$$

where:

$\dot{m}$ is the mass flow rate of outdoor air in kg/s,
h_x is the specific enthalpy of extract air, which is considered as representative of the average indoor air,
h_o is the specific enthalpy of outdoor air,
η_G is the global efficiency of the heat recovery system.

This global efficiency, η_G, is the efficiency of the whole system, including of the ventilated building and its ventilation equipment. It should not be confused with the nominal efficiency of the heat recovery unit itself, ε_{HR}. This efficiency, defined in 'Heat exchange efficiency', above, is measured at the factory with balanced intake and exhaust airflow rates ($\dot{m}_{re} = \dot{m}_{rs}$) and is:

$$\varepsilon_{HR} = \frac{h_{rs} - h_i}{h_x - h_o} = \frac{h_x - h_{re}}{h_x - h_o} \simeq \frac{\theta_x - \theta_{re}}{\theta_x - \theta_o} \tag{5.16}$$

where the signification of subscripts can be seen in Figure 5.8, and h are specific enthalpies of the air in J/kg. As a first approximation, only sensible heat is considered, and the temperatures at the same locations can be used. As shown below, this replacement leads to optimistic results when the air handling unit has parasitic recirculation or when the building has infiltration or exfiltration.

Global heat recovery efficiency

Without heat recovery, the heat loss of the building, Φ_L, resulting from these airflow rates is the sum of extract heat flow and exfiltration heat loss, or the heat necessary to bring outdoor air to indoor climate conditions:

$$\Phi_L = (\dot{m}_x + \dot{m}_{\text{exf}})(h_x - h_o) = (\dot{m}_s + \dot{m}_{\text{inf}})(h_x - h_o) \tag{5.17}$$

The heat recovered by the exchanger is:

$$\Phi_R = \dot{m}_{re}(h_x - h_{re}) = \dot{m}_{rs}(h_{rs} - h_i) \tag{5.18}$$

since, in a first approximation, all the heat taken from extract air is given to supply air. The global heat recovery efficiency of the system is then:

$$\eta_G = \frac{\Phi_R}{\Phi_L} = \frac{\dot{m}_{re}(h_x - h_{re})}{(\dot{m}_x + \dot{m}_{\text{exf}})(h_x - h_o)} = \frac{\dot{m}_{re}}{(\dot{m}_x + \dot{m}_{\text{exf}})}\varepsilon_{HR} \tag{5.19}$$

It can readily be seen that this global efficiency is not equal to the nominal efficiency of the heat recovery system, ε_{HR}. An expression giving η_G as a function of the outdoor airflow, exfiltration and recirculation rates can be derived from Equation 5.19 by taking account of mass conservation at the nodes of the system.

We have mentioned above the following recirculation rates:

External $\qquad R_e = \dfrac{\dot{m}_i - \dot{m}_o}{\dot{m}_e} = \dfrac{\dot{m}_e - \dot{m}_a}{\dot{m}_e}$ (5.20)

Inlet to exhaust $\qquad R_{ie} = \dfrac{\dot{m}_i - \dot{m}_{rs}}{\dot{m}_i} = \dfrac{\dot{m}_e - \dot{m}_{re}}{\dot{m}_i}$ (5.21)

Extract to supply $\qquad R_{xs} = \dfrac{\dot{m}_s - \dot{m}_{rs}}{\dot{m}_x} = \dfrac{\dot{m}_x - \dot{m}_{re}}{\dot{m}_x}$ (5.22)

The mass flow balance for the whole building is:

$$\dot{m}_a + \dot{m}_{exf} = \dot{m}_o + \dot{m}_{inf}$$ (5.23)

Combining this equation with the definition of the external recirculation rate, we get:

$$\dot{m}_e = \frac{1}{1 - R_e}(\dot{m}_o + \dot{m}_{inf} - \dot{m}_{exf})$$ (5.24)

Then, writing the mass flow rate balance at node 1 (see Figure 5.8), we get:

$$\dot{m}_i = \dot{m}_o + R_e \dot{m}_e = \frac{\dot{m}_o + R_e(\dot{m}_{inf} - \dot{m}_{exf})}{1 - R_e}$$ (5.25)

The mass balance at node 2 gives:

$$\dot{m}_{rs} = (1 - R_{ie})\dot{m}_i = \frac{(1 - R_{ie})}{(1 - R_e)}[\dot{m}_o + R_e(\dot{m}_{inf} - \dot{m}_{enf})]$$ (5.26)

From mass balances at nodes 3 and 4:

$$\dot{m}_s = \dot{m}_{rs} + R_{xs}\dot{m}_x$$ (5.27)

and

$$\dot{m}_s = \dot{m}_x + \dot{m}_{exf} - \dot{m}_{inf}$$ (5.28)

we get

$$\begin{aligned} \dot{m}_x &= \frac{1}{1 - R_{xs}}[\dot{m}_{rs} + \dot{m}_{inf} - \dot{m}_{exf}] \\ &= \frac{\dot{m}_o(1 - R_{ie}) + (1 - R_e R_{ie})(\dot{m}_{inf} - \dot{m}_{exf})}{(1 - R_{xs})(1 - R_e)} \end{aligned}$$ (5.29)

Mass balance at node 5 gives:

$$\dot{m}_{re} = \dot{m}_x(1 - R_{xs})$$ (5.30)

Therefore:

$$\eta_G = \frac{\dot{m}_x(1 - R_{xs})}{\dot{m}_x + \dot{m}_{exf}}\varepsilon_{HR} = \eta_x \eta_{re}\varepsilon_{HR}$$ (5.31)

where:

$$\eta_x = \frac{\dot{m}_x}{\dot{m}_x + \dot{m}_{\text{exf}}} \tag{5.32}$$

is the extraction efficiency, i.e. that part of the air leaving the ventilated volume, which is extracted through the air handling unit, and

$$\eta_{re} = 1 - R_{xs} = \frac{\dot{m}_{re}}{\dot{m}_x} \tag{5.33}$$

is the air recovery efficiency, or that part of the extract air that passes through the heat recovery unit.

Looking at Equation 5.32, it seems at first glance that the global heat recovery efficiency depends only on extract and exfiltration airflow rates. However, the purpose of ventilation is to provide fresh, outdoor air in the ventilated volume. Let us see how Equation 5.32 is changed when the fresh airflow rate taken at inlet grille is used as a reference.

Fresh air entering the air handling unit is $\dot{m}_o$. Because of external recirculation, this air is mixed with exhaust air into the inlet duct. A part, R_{ie}, of this mix is recirculated to the exhaust duct. All the fresh, outdoor air that enters the building through the air handling unit is found in $\dot{m}_{rs}$, which is, from the definition of R_{ie} and using Equation 5.25:

$$\dot{m}_{rs} = (1 - R_{ie})\dot{m}_i = (1 - R_{ie})(\dot{m}_o + R_e \dot{m}_e) \tag{5.34}$$

Since $\dot{m}_e$ is no longer fresh, the only part of $\dot{m}_{rs}$ that is fresh is $\dot{m}_o(1 - R_{ie})$. Therefore the total fresh airflow rate entering the ventilated space in building is:

$$\dot{m} = \dot{m}_o(1 - R_{ie}) + \dot{m}_{\text{inf}} \tag{5.35}$$

which means that

$$\dot{m}_o(1 - R_{ie}) = \dot{m} - \dot{m}_{\text{inf}} \tag{5.36}$$

replacing in Equation 5.32 $\dot{m}_x$ by its value given by Equation 5.29, and taking into account the above relation, gives finally:

$$\eta_G = \frac{\lfloor 1 - \gamma_{\text{exf}} - R_e R_{ie}(\gamma_{\text{inf}} - \gamma_{\text{exf}})\rfloor(1 - R_{xs})}{1 - R_e R_{ie}(\gamma_{\text{inf}} - \gamma_{\text{exf}}) - \gamma_{\text{exf}}[R_e + R_{xs}(1 - R_e)]}\varepsilon_{HR} \tag{5.37}$$

where

$$\gamma_{\text{inf}} = \frac{\dot{m}_{\text{inf}}}{\dot{m}} \quad \text{and} \quad \gamma_{\text{exf}} = \frac{\dot{m}_{\text{exf}}}{\dot{m}} \tag{5.38}$$

are respectively the infiltration and exfiltration ratios. In other terms, the extraction efficiency is:

$$\eta_x = \frac{1 - \gamma_{\text{exf}} - R_e R_{ie}(\gamma_{\text{inf}} - \gamma_{\text{exf}})}{1 - R_e R_{ie}(\gamma_{\text{inf}} - \gamma_{\text{exf}}) - \gamma_{\text{exf}}[R_e + R_{xs}(1 - R_e)]} \tag{5.39}$$

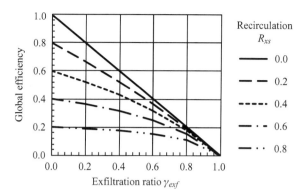

Figure 5.9 *Relative decrease of global heat recovery efficiency as a function of exfiltration ratio γ_{exf} and internal recirculation rate R_{xs}*
Source: Roulet *et al.*, 2001.

which depends on all parasitic airflow rates. When there is no external recirculation ($R_e = 0$), Equation 5.37 simplifies to:

$$\eta_G = \frac{(1 - \gamma_{\text{exf}})(1 - R_{xs})}{1 - R_{xs}\gamma_{\text{exf}}} \varepsilon_{HR} = \eta_x \varepsilon_{HR} \qquad (5.40)$$

and infiltration has no effect. In this case, exfiltration through the envelope and internal recirculation from extract to supply ducts have the same effect, since both drive air away from the heat recovery device. The extraction efficiency in Equation 5.40 is illustrated in Figure 5.9, which indeed represents the relative reduction of heat recovery resulting from exfiltration and internal recirculation.

Global efficiency η_G equals the effectiveness ε_{HR} only if there is no exfiltration, and neither external nor extract-to-supply recirculation. Otherwise, η_G is smaller than ε_{HR}.

The inlet to exhaust recirculation, as well as the infiltration ratio, have only a small effect on heat recovery efficiency, but reduce the amount of fresh air supplied by the unit to the ventilated space. In order to get the same amount of fresh air, the supply airflow rate should be increased. Fresh air efficiency can be defined by:

$$\eta_o = \frac{\dot{m} - \dot{m}_{\text{inf}}}{\dot{m}_s} = \frac{\dot{m}_o(1 - R_{ie})}{\dot{m}_s} \qquad (5.41)$$

This recirculation obviously results in an increased consumption of electric energy for the fans, which is approximately proportional to the cube of the airflow rate, without delivering more fresh air. However, such parasitic recirculation is often not noticed, and hence can lead to an undiscovered reduction of indoor air quality.

Net energy saving and performance index

Heat recovery systems recover thermal energy but use electric energy for the fans. The net energy saving should therefore take into account the primary energy needed to produce electricity and the fact that the losses of the fans heat the air. The net energy saving per cubic metre of supplied outdoor air ($SNES$ in Wh/m^3) averaged over a heating period is:

$$SNES = \rho_o \frac{\eta_G \Phi_L + \Phi_{\text{fan}}(f_r - f_p)}{\dot{m}} \tag{5.42}$$

where:

$\Phi_L = \dot{m}c(\bar{\theta}_x - \bar{\theta}_o)$ is the ventilation heat loss, based on average internal and external temperature during the heating season;

f_r \qquad is the part of the fan power recovered as heat in the supply air. This factor f_r is close to one for supply fans and zero for exhaust fans;

f_p \qquad is a production factor, accounting for the fact that the production of 1 kWh of electric energy requires much more primary energy.

A net gain in thermal or primary energy is achieved by the heat recovery system only when SNES is positive. Otherwise the system even wastes energy.

By analogy with heat pumps, a coefficient of performance, COP, is defined by the ratio of recovered heating power and used electric power:

$$COP = \frac{\eta_G \Phi_L + f_r \Phi_{\text{fan}}}{\Phi_{\text{fan}}} \tag{5.43}$$

This COP is defined without taking account of the production factor, f_p, as is usually the case for heat pumps.

Examples of application

Airflow rates and heat exchanger efficiencies were measured in ten large units and three small, wall-mounted room ventilation units. The main characteristics of these units are summarized in Table 5.2.

Recirculation ratios and efficiencies measured in these units are given in Table 5.3 and illustrated in Figure 5.10. The specific net energy saving ($SNES$) and COP are calculated with a 16 K indoor–outdoor average temperature difference during 210 days, a recovery factor for fans, $f_r = 0.5$ (taking account that there are two fans in these units, one of them in the supply duct) and a production factor, $f_p = 3.55$, which is the average for low-voltage electricity in Europe according to Frischtknecht *et al.* (1994).

Major leakages have been observed in several buildings. In four of them, infiltration represents a significant part of the outdoor air, and in four of them, most of the air leaves the building through the envelope instead of passing the heat recovery unit. Significant internal recirculation is observed

Table 5.2 *Measured airflow rates with experimental uncertainty band (when available), total and specific fan power in audited units*

Unit	Outdoor air	Supply air	Extract air	Exhaust air	W	Wh/m³
		Airflow rates (m³/h)			Fan power	
I	1900 ± 100	2070 ± 70	1790 ± 40	1600 ± 200	990	0.27
2	2530 ± 80	2900 ± 200	1860 ± 50	1500 ± 200	850	0.19
3	2380 ± 70	2480 ± 70	1930 ± 40	1830 ± 50	1800	0.42
4	2200 ± 300	3400 ± 100	3240 ± 90	2000 ± 2000	1800	0.33
5	5000 ± 200	5400 ± 100	6000 ± 700	5500 ± 700	3710	0.34
6	15,000 ± 2000	16,400 ± 700	11,000 ± 1000	10,000 ± 3000	11,800	0.45
7	11,000 ± 400	11,600 ± 200	10,000 ± 300	9500 ± 900	8180	0.39
8	16,000 ± 1000	17,400 ± 700	13,400 ± 600	12,000 ± 2000	9760	0.33
9	9000 ± 1000	10,000 ± 2000	1970 ± 90	1000 ± 3000	3800	0.35
10	14,300 ± 600	16,200 ± 400	3420 ± 70	1000 ± 1000	7970	0.45
a	25	36	34	24	13	0.22
b	42	75	74	41	27	0.24
c	74	87	87	74	32	0.20

Table 5.3 *Outdoor air efficiency, η_o, exfiltration and infiltration ratios γ_{exf} and γ_{inf}, external and internal recirculation rates R_e, R_{xs} and R_{ie}, heat recovery effectiveness ε_{HR}, global heat recovery efficiency η_G, specific net energy saving, SNES in Wh/m^3, and coefficient of performance, COP, of audited air handling units*

Unit	η_o	γ_{exf}	γ_{inf}	R_e	R_{xs}	R_{ie}	η_x	ε_{HR}	η_G	SNES	COP
I	97%	16%	0%	6%	7%	0%	86%	70%	56%	1.55	6.5
2	92%	47%	9%	20%	5%	0%	59%	70%	39%	1.35	8.0
3	100%	29%	7%	0%	5%	0%	72%	90%	62%	1.18	5.2
4	68%	77%	76%	55%	1%	0%	31%	30%	9%	−0.05	3.3
5	98%	8%	17%	0%	7%	0%	92%	80%	69%	1.92	6.7
6	97%	43%	8%	0%	6%	0%	61%	90%	52%	0.69	4.5
7	100%	14%	0%	4%	2%	0%	87%	80%	68%	1.45	5.5
8	97%	25%	0%	0%	0%	0%	77%	70%	54%	1.17	5.5
9	95%	97%	49%	0%	0%	0%	10%	50%	5%	−0.37	1.8
10	93%	91%	18%	100%	6%	0%	18%	50%	8%	−0.92	1.5
a	74%	8%	0%	0%	33%	0%	94%	63%	40%	1.37	6.2
b	57%	2%	0%	0%	44%	4%	99%	80%	44%	2.21	6.8
c	68%	0%	0%	0%	39%	25%	100%	90%	55%	2.69	8.2

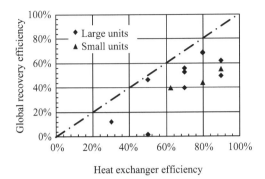

Figure 5.10 *Global heat recovery efficiency versus nominal heat exchanger effectiveness measured in several units*

in the three small units, and external recirculation above 20 per cent is measured in three large units. These leakages significantly affect heat recovery efficiencies, which drop from nominal values of between 50 and 90 per cent down to actual values ranging between 5 and 69 per cent. On average, the nominal heat recovery effectiveness ε_{HR} is 70 per cent, but the global, real efficiency is only 43 per cent. In the best case, 80 per cent heat recovery effectiveness is reduced by 15 per cent down to a 69 per cent real efficiency.

Only 8 units out of 13 have a net energy saving larger than 1 Wh/m, as shown in Figure 5.11. Note that 1 Wh allows heating one cubic metre of air by about 3°C. Negative specific net energy savings are observed in three units, where the heat recovery uses more energy than it saves! The coefficient of performance of good units can be much larger than those of a heat pump used for heating buildings, but is rather small in three units. A *COP* of less than 2.5 indicates that the heat recovery is less efficient than heating the air with a gas boiler with 75 per cent efficiency (Ruyssevelt, 1987).

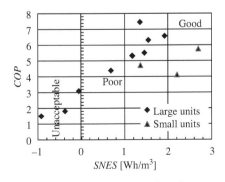

Figure 5.11 *Seasonal average coefficient of performance and specific net energy saving of the tested units*

Best net energy savings in large units (7 and 8 in Table 5.2) are 80,000–90,000 kWh per winter season, but unit 10 actually wastes as much energy. Small units (a, b and c) save between 80 kWh and 350 kWh during an entire season. From an energy and economic lifetime analysis perspective, such units are disadvantageous.

Note that these results are obtained when the heat recovery is functioning. Annual average efficiency may even be less due to reduced operation time (Drost, 1993).

Energy for ventilation

The energy to move the air is the product of a force by a displacement. The force is the pressure, Δp, exerted on the section area, A, of the duct, and the displacement is the path, l, of the air during a time interval, Δt. But $A \cdot l$ is the volume of air displaced during this time interval. The energy to move a volume V of air is hence:

$$E_m = \Delta p A \cdot l = \Delta p V \tag{5.44}$$

Taking a time derivative of the above equation provides the mechanical power, Φ_m, needed to get an airflow rate, q:

$$\Phi_m = \frac{dE_m}{dt} = \Delta p \frac{dV}{dt} = \Delta p q \tag{5.45}$$

The mechanical power delivered by a fan is the product of the volume airflow rate, Q, delivered by the fan, and the pressure differential, Δp, across the fan. The mechanical power required to move the air through a ductwork is also the product of the volume airflow rate through the ductwork, and the pressure difference between the main supply and main exhaust ducts. Since the pressure difference is proportional to the square of the airflow rate, the mechanical power for ensuring a given airflow rate into a ductwork is proportional to the cube of the airflow rate! Increasing the airflow rate in a room by 10 per cent requires 33 per cent more fan power and doubling the airflow rate requires a power eight times larger if the ductwork is not adapted to this new airflow rate.

Why check fan power and related quantities?

The electrical energy needed to move the air depends on the properties of the air distribution system and of the fan. For a given nominal power, efficiencies varying by a factor two or more were measured (see 'Examples of application', below). Assessing the fan efficiency and the specific power (in Joules or Watt-hours per cubic metre of transported air) is part of a comprehensive energy diagnosis of a mechanical ventilation system.

Poor fan efficiency not only wastes expensive electric energy, but also hinders efficient cooling. The cooling power of the air blown by the fan is:

$$\Phi_{\text{cool}} = cQ\,\Delta\theta = c\rho q\,\Delta\theta \tag{5.46}$$

where:

ρ is the density of air,
c is the heat capacity of air,
$\Delta\theta$ is the temperature difference between exhaust air and supply air.

The kinetic energy given to the air by the fan is, sooner or later, degraded into heat by viscosity and friction on the surfaces of ducts, room walls and furniture. The kinetic energy of the air leaving the room to the outside is very small when compared to that of the air just after passing through the fan, especially in units with large recirculation ratios. Since the fan motor is in the airflow, its heat loss is also delivered to the air. Therefore, nearly all the energy given to the fan ends as heat in the indoor air. This corresponds to a heating power equal to the electric power consumed by the fan motor, Φ_e. Hence:

$$\Phi_{\text{heat}} = \Phi_e = \frac{q\,\Delta p}{\eta_f} \tag{5.47}$$

For air conditioning, the heating power should be small when compared to the cooling power. Therefore, the ratio:

$$\frac{\Phi_{\text{cool}}}{\Phi_{\text{heat}}} = \eta_f \frac{\rho c\,\Delta\theta}{\Delta p} \tag{5.48}$$

should be as large as possible. This means that the fan efficiency should be as close as possible to one (or 100 per cent). In addition, the pressure differential should be as small as possible.

Another way to look at this issue is to calculate the air temperature increase resulting from heat loss:

$$\Delta\theta_{\text{heat}} = \frac{\Phi_e}{\rho c q} = \frac{\Delta p}{\eta_f \rho c} \tag{5.49}$$

This should be as small as possible, so again, the fan efficiency should be large and the pressure differential should be at a minimum.

The energy losses of fans are shared between the elements of the chain linking the electrical network to the aeraulic ductwork (see Figure 5.12). In this chain, the fan is often the worst culprit. It is not, however, simple to assess the efficiency of each element, and we will concentrate on the measurement of the efficiency of the whole chain, by measuring on the one hand the

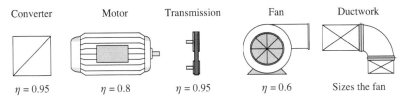

Figure 5.12 *Approximate figures for the efficiencies of various elements needed to move the air in the ductwork*

consumption of electrical energy by the fan motor, and on the other hand the kinetic energy given to the air in duct.

The fan efficiency is the ratio of useful power, Φ_m, to the electrical power consumed by the fan motor, Φ_e:

$$\eta_f = \frac{\Phi_m}{\Phi_e} = \frac{q\,\Delta p}{\Phi_e} \tag{5.50}$$

Measuring the airflow rate, q, and the pressure differential, Δp, across the fan provides the kinetic power of the air, and measuring in addition the electric power used by the fan allows for the assessment of fan efficiency.

The uncertainty band resulting from uncertainties on the measured quantities is:

$$\delta\eta_f = \eta_f \sqrt{\left[\frac{\delta(q)}{q}\right]^2 + \left[\frac{\delta(\Delta p)}{\Delta p}\right]^2 + \left[\frac{\delta\Phi_e}{\Phi_e}\right]^2} \tag{5.51}$$

Measurement of airflow rate

Airflow rates through both supply and exhaust fans can be assessed by tracer gas measurements, as described in Chapter 2, 'Airflow measurements at ventilation grilles'. Depending on the units used when interpreting the results, these measurements may provide either volume airflow rates, q_v, or mass airflow rates, Q_m. These are related by:

$$Q_m = \rho q \tag{5.52}$$

where ρ is for the density of the air, which can be calculated by:

$$\rho = \frac{\bar{M}p}{RT} \cong 3.46 \cdot 10^{-3}\,\frac{p[\mathrm{Pa}]}{T[K]}\,[\mathrm{kg/m^3}] \tag{5.53}$$

where:

p is the atmospheric pressure (average 101,300 Pa at sea level),
T is the absolute temperature,
$\bar{M}$ is the average molar mass of the air mixture (about 28.8 g/mole),
R is the molar gas constant $= 8.31396\,\mathrm{J/(moleK)}$.

Then, if mass airflow rates are taken from tracer gas measurements, the airflow through the fans should first be converted into volume airflow rate.

Airflow rate through the fan can also be assessed from the fan speed, the pressure differential across the fan, and the fan characteristics – provided by the factory – which give the airflow rate from the fan speed and pressure differential.

Measurement of pressure differences

The pressure differential is measured with a differential manometer with a range of 200–500 Pa (20–50 mm water column).

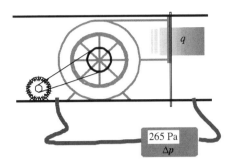

Figure 5.13 *Installation of the differential manometer to measure the pressure differential across the fan*

The two ports of this manometer are connected to pressure taps located on both sides of the fan (see Figure 5.13). Care should be taken to avoid too much dynamic pressure on these taps. It is advisable to install the pressure taps perpendicular to the airflow, preferably close to the air duct wall, and at locations where the air velocity is about the same on both sides of the fan so that the dynamic pressure, if any, is the same on both sides. If the pressure varies significantly when moving one of the pressure taps, this means that the dynamic pressure has an effect.

On most air handling units, a differential pressure switch is installed to check the function of the fan. This switch is connected by two pipes to taps installed in the ducts before and after the fan. These taps can be used to connect the differential manometer, but the safety switch should be either disabled or short-circuited. Otherwise, the fan motor will stop as soon as the pipes of the pressure switch are disconnected.

Measurement of electric power

The electric power used by the fan motor is measured with a wattmeter, which should be wired to the fan according to Figure 5.14. This instrument measures

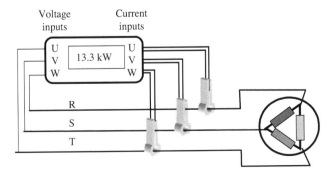

Figure 5.14 *Schematics of electric power measurement on a three-phase motor*

simultaneously the r.m.s. voltage, U, between phase and neutral point, the r.m.s. current, I, running into each motor coil and the phase shift, φ, between voltage and current. The power is calculated by:

$$\Phi_e = \sum_{j=1}^{3} U_j I_j \cos(\varphi_j) \tag{5.54}$$

the sum being on all three phases.

In very small air handling units, the fan motor may be single phase, and

$$\Phi_e = U I \cos(\varphi) \tag{5.55}$$

To measure the current an ampere-meter must be installed in the circuit. An easy way is to use clamp-on ampere-meters. A measuring clamp is installed around each wire leading to the motor. This clamp contains a transformer that gives a current proportional to the current running through the closed clamp. Care should be taken not to install the clamp around the two-wire or four-wire cable. The measured current will be zero in this case, whatever the power used by the motor.

In some air handling units, the fan is controlled by a variable frequency controller (see Figure 5.15). Such devices are often equipped with a screen

Figure 5.15 *Front panel of the variable frequency controller*

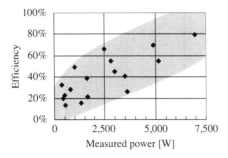

Figure 5.16 *Fan efficiencies as a function of actual fan motor power*

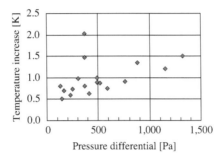

Figure 5.17 *Air temperature increase as a function of pressure differential across the fan*

on which the frequency, the voltage, the current and the fan motor power can be displayed.

Examples of application

Fan efficiencies were measured on several fans of various units. Figure 5.16 represents the measured fan efficiencies versus their measured used electric power. It shows a general improvement of the fan efficiency when fan motor power increases. However, the dispersion is huge and large differences can be observed for each power class. For example, efficiencies range from 30 per cent to more than 60 per cent for 3 kW fans, and from 10 per cent to 35 per cent for small fans. Figure 5.17 illustrates the fact mentioned above (see 'Why check fan power and related quantities?') that the air temperature increases when the pressure differential is large. The dispersion results from variations in fan efficiencies.

Energy effects of indoor air quality measures

In the 1970s, after the oil crises, measures were hastily taken in many buildings to reduce their energy use. These measures were planned with only two objectives: energy efficiency and return on investment, without taking care of indoor environment quality and health or paying attention to possible damages to buildings. If a decrease of thermal comfort was implicitly accepted, cases of mould growth, increased indoor pollution and health hazards were not expected but often observed. Since then, the idea that saving energy in buildings decreases the indoor environment quality still prevails.

Of course, some energy conservation opportunities such as low internal temperature or too low ventilation rates may degrade the indoor environment. These should therefore either be avoided, or accepted only in case of emergency and for a limited period of time.

Some other energy saving measures should be used only in conjunction with others. For example, retrofitting windows in poorly insulated dwellings leads to a risk of mould growth, and improving the envelope airtightness without taking care of ensuring and controlling a minimum ventilation rate may decrease indoor air quality.

Table 5.4 lists, in the first column, various uses of energy in buildings. Known ways to save energy are presented in the second column, and effects of these energy saving measures on comfort or indoor environment quality are presented in the third column. It can readily be seen that there are many cases where energy saving measures, when well designed and executed, improve indoor environment quality.

Several recommendations, resulting from experience and recent surveys performed within European projects (Bluyssen *et al.*, 1995; Roulet *et al.*, 2005) are given below.

Method

The method used to scientifically support these recommendations is described in detail in Jaboyedoff *et al.* (2004). A typical office building equipped with full air conditioning and cold ceiling, including heat recovery, was simulated using an appropriate computer program. The heating, ventilation and air conditioning (HVAC) system is shown in Figure 5.18.

The three-storey building is of heavy construction and well insulated with low-e, clear glazing. The office rooms are oriented south with 53 per cent glazed area. The internal temperature can be controlled using either air conditioning or hydronic heating and cooling.

Numerous simulations were performed for different climates: northern (Oslo), central (London and Zürich) and southern Europe (Rome), and the following variants were calculated:

- outdoor air supply: 15–50 m^3/hour and per person;
- relative humidity of supply air to room: 50 per cent and no humidification;
- efficiency of heat recovery: 0 (no heat recovery), 0.75 and 0.85;
- infiltration: 0.5 and 1.0 air changes per hour;
- set point for cooling: 24°C, 28°C and no cooling;
- ventilation: 24 hours a day or during working hours only (7 am to 7 pm);
- natural ventilation – using windows – instead of mechanical ventilation.

Table 5.4 *Uses of energy in buildings, energy saving measures and their effects on indoor environment quality*

Energy use	Energy saving measures	Impact on indoor environment
Compensation of transmission heat loss in winter	Better, thicker insulation IR reflective by low emissivity coated	Improves comfort and health by preventing mould growth
Compensation of ventilation heat loss in winter	Lower ventilation rate	Needs a reduction of indoor pollution sources to maintain indoor air quality
	Limit the ventilation rate to the required level Use heat recovery on exhaust air	Less draughts, less noise Generally improves indoor air quality in winter
Winter heating in general	Improve solar gains with larger, well-oriented windows Improve the use of gains by better insulation and good thermal inertia	If windows are poor: cold surfaces Overheating if poor solar protection If well designed: good visual contact with outdoor environment, excellent summer and winter comfort
Elimination of heat gains during warm season	Use passive or 'free' cooling Use efficient, well-commissioned and maintained cooling systems Higher internal temperature	Very comfortable in appropriate climates and buildings Better indoor air quality and comfort Should be kept within comfort zone
Internal temperature control	Comfortable set-point temperature, improved control	Avoids over- and under-heating
Ventilation (moving the air)	Natural ventilation Reduce airflow rate	Best where applicable Possible only where overventilated
	Increase duct size Efficient fans	Less noise Less noise
Humidification	Switch it off	No health effect in most cases
Lighting	Use daylighting	Comfortable light, with limited heat gains when well controlled
	Use efficient artificial lighting	Comfort depends on the quality of light. Reduced heat load

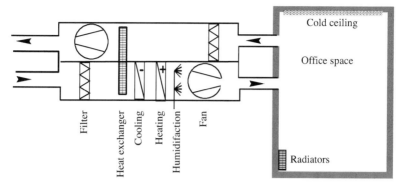

Figure 5.18 *The HVAC system in the simulated building*

For each variant, the effect of the following changes in design and operation was simulated:

- with 50 per cent recirculation instead of no recirculation;
- with or without heating or cooling supply air;
- effect of ventilation strategies on heating demand;
- effect of airtightness on heating demand;
- with and without a droplet catcher with 20 Pa pressure drop downwind of the cooling coil;
- with an improved filter with 150 Pa pressure drop;
- changing used filters at a pressure drop of 180 Pa instead of 250 Pa;
- with rotating or flat plate heat exchanger: efficiency 0.85 and 0.75;
- effect of increase of pressure difference through ductwork: 1600 Pa versus 1000 Pa.

Simulation results

The main results of these simulations are summarized below. These conclusions, in particular the numerical values, are valid for the building and the system simulated. However, the order of magnitude and general tendencies are likely to be valid for more general situations.

Recirculation

The electric energy used by cooling and fans decreases by about 40 per cent (27 per cent for Rome, 43 per cent for Zürich and 50 per cent for Oslo) if no recirculation is used, compared to 50 per cent recirculation. No heat recovery was used in these cases.

Heating

The energy use for heating mainly depends on climate and internal gains. Except for heat recovery and time schedule of operation (working hours/24 hours per day), ventilation strategies have a minor influence on heating energy demand. The air may be either heated by coils in the supply air or by radiators in the room. The energy use for heating does not change significantly in all climates.

The tightness of the building envelope has a large influence, up to a factor of two, on the heating energy need. When high infiltration occurs, humidity is also reduced in winter.

Cooling

Cold ceilings are more effective than air conditioning. For the same airflow rate and same comfort conditions, more energy was required for cooling using air conditioning than with the hydronic cooling ceiling.

Lowering the set point for cooling from 26°C to 24°C causes an augmentation of the cooling demand of the zones by a factor of three to eight, depending on the geographic location.

A droplet catcher downwind of a cooling coil has a negligible effect on energy demand, but may be essential to avoid humidifying downstream filters or acoustic dampers, changing them to biotopes.

Filters

Using a two-stage filter system instead of an old F7 filter leads to an increase in electric power use for fans of 10–15 per cent, depending on the pressure difference over the system.

Again, depending on the pressure difference over the system, earlier replacement of a filter results in a decrease in fan power consumption by 2–3 per cent.

Humidification

In the northern (Oslo) and central locations (London and Zürich), humidifying the supply air at 30 per cent minimum relative humidity requires about 20–25 per cent more energy for ventilation than without humidification. In the southern climate (Rome), the increase is only 3 per cent, mainly because humidification is seldom required. In all climates, humidifying the supply air at 30 per cent increases the total heating energy need by 5–10 per cent, while this need almost doubles if the relative humidity is set at 50 per cent.

Heat recovery

Without heat recovery, the heating energy use for ventilation is 70–140 per cent more than with medium efficiency (50 per cent) heat recovery.

Heat recovery with high efficiency (75 per cent) – such as those achieved by well-installed rotating heat exchangers in airtight buildings – leads to a reduction of the heating energy demand for ventilation by about 30 per cent, compared to medium efficiency (50 per cent) heat recovery. That means that the 3 per cent reduction in efficiency caused by installing a purging sector in a rotating heat exchanger (see Chapter 6, 'Rotating heat exchangers'), has a negligible effect on energy demand.

Infiltration or exfiltration through a leaky building envelope strongly reduces the efficiency of heat recovery (see 'Effect of leakages and shortcuts on heat recovery', above). With a heat recovery efficiency of 75 per cent, the heating energy demand for cold and mild climates (Oslo, Zürich, London) is approximately 3–5 per cent higher with an efficiency of 0.75 than with 0.85. For warm climates (Rome), this number is approximately 15 per cent higher, however, with low absolute values.

Ductwork

An increase of pressure difference from 1000–1600 Pa, caused by air velocity, length, curves, duct wall smoothness and deposits in the ducts, leads to an increase in electric power use of 60 per cent. The increase in total electric power depends on the geographic location and ranges from 25–55 per cent.

Notes

1 Natural ventilation can be controlled by installing (automatically or manually) adjustable vents in an airtight building envelope.
2 These ratios take not only density and heat capacity into account, but also practical temperatures.

6

Contaminants in
Air Handling Units

The purpose of mechanical ventilation systems is to supply appropriate amounts of clean air and to evacuate vitiated air. However, in field audits it was seen that ventilation systems often host contaminant sources and are, in the worst cases, the main source of air pollution in buildings (Fanger, 1988; Bluyssen *et al.*, 1995, 2000b). Components in the mechanical ventilation system may considerably pollute the passing air. The main sources and reasons for pollution in a ventilation system vary considerably depending on the type of construction, use and maintenance of the system. This chapter summarizes the results of these field audits, and proposes methods to detect the sources of contaminants and strategies to avoid these.

Filters

Filters are one of the main sources of sensory pollution in ventilation systems (Bluyssen *et al.*, 2000a). Some new filters may also influence the perceived air quality negatively. The filter material has a significant influence on the starting pollution effect of new filters (see Figure 6.1).

When filters get older, i.e. are in use for some time, the emission of odours first decreases, but increases again later, when the filter gets loaded. The reason for this emission after the filter is in use for some time remains unclear, however.

Micro-organisms may not be the only pollution source on a filter, but it is important to keep filters dry, since wet media filters are perfect supports for microbial growth and microbes may also emit dangerous pollutants and bad odours. Filters may be moistened either by snow, rain or fog entering the outdoor air inlet, or by water droplets spread by some humidifiers or found in airflows downstream of the cooling coils.

Environmental conditions such as airflow (amount or intermittent/continuous) and temperature do not have a significant influence on the pollution of downwind air.

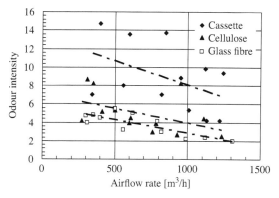

Figure 6.1 *Olfactive pollution of various new filters as a function of airflow rate*

Source: Bluyssen *et al.*, 2000a, 2003.

Ducts

The duct material and the manufacturing process has the biggest effect on the perceived air quality (Björkroth *et al.*, 2000). Depending on the machinery used in the manufacturing process, new spiral wound ducts, flexible ducts and other components of the ductwork might contain small residual amounts of processing oil. The oil layer is very thin and invisible, but it emits an annoying odour. Aluminium ducts score the best with respect to perceived air quality. Plastic ducts seem a feasible solution, but some flexible plastic ducts are very smelly.

Oil residues are the dominating sensory pollution source in new ducts. The sensory assessments showed a clear correlation between the total mass of oil residues (average surface density × surface area) and the perceived air pollution (see Figure 6.2).

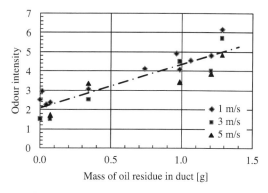

Figure 6.2 *Correlation between odour intensity and the mass of oil residues in the tested ducts*

Source: Björkroth *et al.*, 2000.

The effect of airflow on the perceived air quality from ducts was relatively small and is probably insignificant in normal applications. Increasing the airflow rate in the duct does not, surprisingly, reduce the odour intensity: the additional airflow rate certainly dilutes the evaporated oil but the increased air velocity also evaporates more oil.

Emissions from dust/debris accumulated in the ducts during construction (mostly inorganic substances) seem to be less important. No simple correlation was observed between the amount of accumulated dust and odour emissions. However, the organic dust accumulated during the operation period may produce more severe odour emissions. When dust has accumulated on the inner surface of the ducts, the relative humidity of the air in the ducts has a larger effect on the odour emissions of ducts without oil residues than ducts with oil residues.

Humidifiers

The main reasons for pollution from humidifiers are: disinfecting additions, old water in tanks or dirty tanks, microbiological growth, stagnant water in the tank when the humidifier is off, and desalinization and demineralization devices and agents (Müller *et al.*, 2000).

Humidifiers only pollute the air significantly if the humidifier is not used or maintained in the prescribed way. Investigations make it clear that periodical cleaning of humidifiers and the use of fresh water are paramount for a good air quality.

Under normal conditions, it was found for all humidifiers that airflow has no influence on the odour intensity caused by humidifiers (see Figure 6.3).

A relation was found between perceived air quality and the concentration of bacteria on the inside of the humidifier (see Figure 6.4). The odour intensity increases with increasing number of bacteria. This was not the case for other locations in an HVAC system. A similar correlation could not be found for fungi.

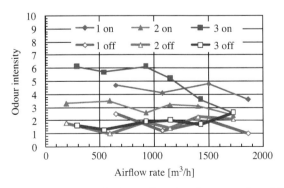

Figure 6.3 *Perceived air quality for the steam humidifier*
Source: Müller *et al.*, 2000.

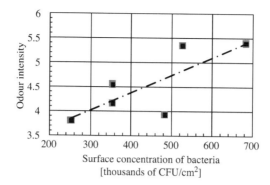

Figure 6.4 *Bacteria concentration at inner surface of a humidifier correlated with the odour intensity*

Note: CFU = colony forming unit.
Source: Müller *et al.*, 2000.

Rotating heat exchangers

Rotating heat exchangers are not themselves sources of contaminants, but they may transfer contaminants from exhaust to supply air with entrained air, and through possible leakage around the wheel at the separation wall. Leakage from exhaust to supply was measured by the author in several units, and found to be negligible in most cases (see Chapter 5, 'Leakage through heat exchangers').

A part of the extract air is indeed entrained to the supply duct by the rotation of the wheel, as shown in Figure 6.5. All the exhaust air contained in a sector of the wheel is entrained back into the supply air.

This is avoided by installing a purging sector, which returns this vitiated air back to the exhaust duct (see Figure 6.6). This chamber covers a sector of about 5°, in which the outdoor air passes through the wheel, makes a 180° turn in the purging chamber, passes back in the wheel and finally leaves the air handling unit by the exhaust air duct. This cleans the wheel from contaminants accumulated when passing in the extract air, before entering the outdoor air. Note that this device functions properly only when the sense of rotation of the wheel is

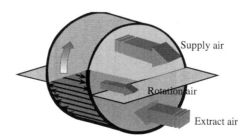

Figure 6.5 *Some extract air is entrained in the supply airflow by the rotation of the wheel*

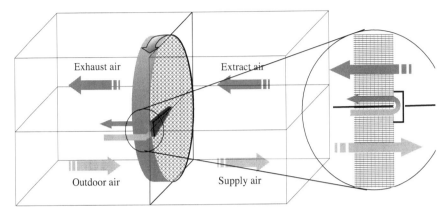

Figure 6.6 *Schematics of the purging sector*

Note: A part of the outdoor air cleans the porous structure and then is sent back to the exhaust air.

such that a sector of it that contains exhaust air passes first through the purging chamber. The author has seen wheels turning the wrong way!

In addition, contaminants can be transferred from exhaust to supply ducts by adsorption–desorption. This was confirmed by measurements with volatile organic compounds (Andersson *et al.*, 1993; Roulet *et al.*, 2000) and perceived air quality (Pejtersen, 1996). For example, measurements performed by the author according to the protocol described in 'Contaminant transport in rotating heat exchangers', below, gave the transfer rates illustrated in Figure 6.7. This figure shows transfer rates with and without a purging sector.

Leakage and entrained air would result in the same recirculation rate for all chemical compounds, this rate being close to zero for the unit giving the results of Figure 6.7. This is obviously not the case.

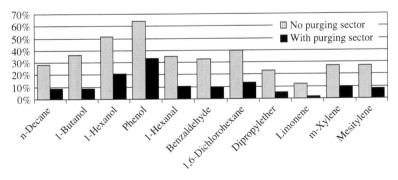

Figure 6.7 *Average VOC recirculation rates measured in the EPFL laboratory unit, with and without a purging sector*

Source: Roulet *et al.*, 2000.

Figure 6.7 shows that certain categories of volatile organic compounds (VOCs) are easily transferred by a sorption transfer mechanism. Among the tested VOCs, those having the highest boiling point were best transferred. The largest transfer rate in a well-installed unit was found for phenol (30 per cent).

Leakage and pollutant transfer can be avoided or at least strongly reduced through proper installation of the wheel, good maintenance of the gasket, proper installation of a purging sector, and by maintenance of a positive pressure differential from supply to exhaust duct at wheel level.

Coils

Laboratory tests (Bluyssen *et al.*, 2003) show that heating and cooling coils without condensing or stagnating water, are components that have small contributions to the overall odour intensity of the air. On the contrary, cooling coils with condensed water in the pans are microbial reservoirs and amplification sites that may be major sources of odours to the inlet air.

Measurement protocols

HVAC systems are in general low sources of measurable chemical pollutants. When searched for, most pollutants are below the detection limits of common analysers, and chemical analyses can be successful only in very polluted systems. They are therefore not discussed here.

No standard procedure exists for microbiological measurements in ventilation systems. The techniques used are air sampling with impactors, gluing airborne microbes (mould, yeast, bacteria) on appropriate culture media, or simply exposing these culture media in open Petri dishes or on films lying or glued on the inner walls of ducts or units. The main problem is ensuring reproducible samples.

Only two methods are presented here: the measurement of sensory pollution and the assessment of contaminant transfer.

Sensory pollution

Principle of the method

Since the nose is the most sensitive instrument to detect pollutants, the measurement protocol to assess the pollution resulting from ventilation systems or components is mainly focused on measuring the sensory pollution effect, evaluated by a trained panel of people (Bluyssen, 1990; Elkhuizen *et al.*, 1995).

A panel of 12–15 subjects is selected and trained to give a value to the odour intensity. To evaluate air quality at a given place, each panel member smells the air – after having refreshed his or her nose in pure, fresh air – and gives a value to the odour intensity. The final value is the average over the panel.

Selecting the panel

The subjects are selected from a group of at least 50 applicants of ages ranging from 18 to approximately 35 years old. There is no restriction on distribution of gender. Participants should abstain from smoking and drinking coffee for at least one hour before any test. Also, they are asked not to use perfume, strong smelling deodorants or make-up, and not to eat garlic or other spicy food the day before the tests and on the day of the tests.

The selection is based on the quantitative assessment of the concentration of a reference gas by smelling. The reference gas is 2-propanone, which is easy to measure and to produce in various concentrations in the air. Passive evaporation creates known concentration of this gas in air, and this air is presented to the human nose at a constant airflow coming out of a so-called PAP meter, which consists of a 3 l jar made of glass covered with a plastic cap, a fan and a diffuser (see Figure 6.8). The cap has two holes; in one of them the fan is placed to suck the air through the jar and to blow it into a glass cone that diffuses the exhausted air. The angle of the cone is at 8°, to avoid mixing with room air. The diameter of the top of the cone is 8 cm, convenient to situate the nose in the middle. The small fan should produce at least 0.9 l/s, several times more than the airflow during inhalation. The person therefore inhales exclusively air from the jar, undiluted by room air.

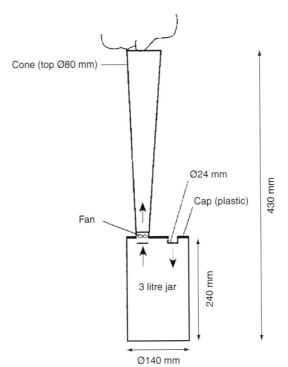

Figure 6.8 *The PAP meter*

Source: Bluyssen, 1990.

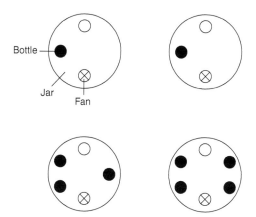

Figure 6.9 *Recommended locations of small bottles in PAP meter*
Source: Bluyssen, 1990.

The 2-propanone gas is evaporated in the PAP by placing one or more 30 ml glass bottles filled with 10 ml of 2-propanone and making different holes in the caps of these bottles. The concentration (in parts per million) of 2-propanone obtained with one small bottle is about three times the diameter of the hole in millimetres. Placing one or more bottles enables the production of different 2-propanone concentrations. The actual 2-propanone concentrations should be measured by a suitable calibrated analyser. The position of the small bottles in the PAP meter is of great importance. Recommended positions are illustrated in Figure 6.9.

The steady-state concentration of 2-propanone in the top of the diffuser depends on the level of the liquid in the small bottles (which is standardized at 10ml), the location of the small bottles in the jar of the PAP meter, the ambient temperature (standardized at 22°C), and the size of the holes in the caps of the small bottles through which the 2-propanone diffuses. To get steady state, it is recommended to condition the whole apparatus and the small bottles with 2-propanone the day before any test. One hour before the test, place the bottles in position in the jar, leave the over-caps off and activate the fan. After 30 minutes a steady-state level with less than 3 per cent variation should be reached.

The space where the sensory panel is trained has to fulfil certain criteria. Preferable is a space that has:

- temperature control;
- 100 per cent outdoor ventilation;
- a filtration unit (for example, active carbon);
- a Teflon layer on walls, floor and ceiling;
- displacement ventilation (from floor to ceiling) or local exhaust.

Acceptable is a space that is empty (no smoking) and has:

- walls, floor and ceiling covered with a Teflon layer or cleaned with a non-smelling agent;

Table 6.1 *PAP values and 2-propanone concentrations in PAP meters used as milestones*

Value	Concentration [ppm]
1 (no odour)	<1
2	5
5	19
10	42
20	87

- mechanical air supply with filtered air;
- mixing ventilation with a certain minimum ventilation rate.

During the tests, the background level of 2-propanone should not be more than 1 ppm.

Five different 2-propanone concentrations generated by five PAP meters are used as milestones for the training. These milestones have PAP values corresponding to the concentrations in Table 6.1 of 2-propanone in the top of the cone.

The equation for calculating the values is:

$$\text{PAP value} = 0.84 + 0.22 \times 2 - \text{propanone concentration (ppm)} \quad (6.1)$$

Once the milestones are calibrated, it is important to keep the same PAP meters with the same bottles at the same locations in the jar, since changing these parameters may change the 2-propanone concentration in the cone.

For selecting the panel, eight additional PAP meters with five concentrations ranging from 0–40 ppm and three with 5–90 ppm are prepared. The applicants are asked to assess these eight different concentrations of 2-propanone using the milestones as the reference. They will be instructed to have at least two inhalations of unpolluted air in between each PAP sniff. The question asked of the applicants is:

> *How strong is the air that you perceive? Give a number on a scale from 1 to 20, but refer this number always to the numbers on the milestones 1, 2, 5, 10 and 20. One is equal to no smell (you perceive nothing), 20 is equal to extreme strong smell.*

The applicants are allowed to go back and forth between the eight different unknown concentrations and the milestones as much as they need, and note the given value for each unknown PAP meter. The dozen subjects with the least sum of differences between given and actual values are selected.

Training procedure

The 12–15 subjects are trained for three to five days in smaller groups of three or four people. Each day, they receive approximately one hour of intensive training. In the first two days, the panel is trained to assess perceived air pollution of concentrations of 2-propanone unknown to them by making comparison with the milestones. On the third, fourth and fifth days, training includes 2-propanone concentrations and other sources of pollution. Since the pollutants have different characters to 2-propanone it is of great importance that the subjects understand that they are exposed to the intensity by comparing the intensity of the milestones. Up to 12 different concentrations of pollutants are presented each training day.

After the evaluation of each unknown concentration of 2-propanone, the panel member is given the correct answer and their performance is discussed with the experiment leader. The panel members write their votes on a form where it is possible to follow their performance during the training. Panel members are submitted to a test similar to the selection test at the end of the third and fifth training days. If they do not succeed, they should either quit the panel or train for two additional days.

Experimental day

On each experimental day, the panel members are retrained for approximately 15–20 minutes per group of three to four people. During this training the panel members are exposed to two or three different concentrations of 2-propanone and two different materials, on which they receive feedback.

Also on each experimental day, the panel members are exposed to six different concentrations of 2-propanone corresponding to the values 1, 3, 7, 12, 16 and 19. The concentrations of 2-propanone should be measured just before the sensory assessments of each group of panel members. These exposures make it possible to compare different sensory panels and to calculate performance factors.

The panel members are placed in a well-ventilated waiting room. During each round of assessments, one by one the subjects assess the intensity of the perceived air pollution of the air sample (from a material in a PAP meter, in a walk-in climate chamber, air from a ventilation system and so on) by making comparison with the intensity of the milestones. The panel members are allowed to go back and forth between the milestones and the polluted air sample. The members write down their assessment on a voting sheet, which is handed to the experiment leader before making the next assessment.

The time between assessments or each panel member should not be less than three minutes. An experienced panel member can assess an air sample within 30–45 seconds. With a panel of 12 subjects the time between assessments for a panel member will be approximately nine minutes, and for a group of four panel members three minutes.

Contaminant transport in rotating heat exchangers

Principle of the method

A heated VOC mix is injected in the extract duct (star in Figure 6.10), in such a way that the VOCs are well evaporated and mixed into the air at location C_4

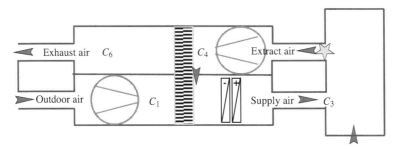

Figure 6.10 *Schematics of an air handling unit showing location of VOC injection and sampling points, C_i, for concentration analysis*
Source: Roulet *et al.*, 2000.

of Figure 6.10. The concentration of these VOCs in the air is analysed at the four locations shown in Figure 6.10. Locations C_6 and C_3 should be far enough from the rotating heat exchanger to ensure a good mixing after sorption and desorption.

If there is no transfer at all, $C_3 = C_1 = C_0$, the outdoor concentration of each compound, and $C_6 = C_4$. If some VOCs are transferred, there should be, at steady state, equilibrium between the amount of VOC taken in extract air and the amount transferred to supply air:

$$Q_e(C_4 - C_6) = Q_s(C_3 - C_1) \tag{6.2}$$

The transfer rate can be calculated as the ratio of the mass flow rate of VOC delivered into supply air, and the mass flow rate of VOC in extract air:

$$R = \frac{Q_s(C_3 - C_1)}{Q_e C_4} = \frac{Q_s(C_3 - C_1)}{Q_e C_6 - Q_s(C_3 - C_1)} \tag{6.3}$$

Injection technique

Injection could be performed in two ways:

1   At a constant flow rate – for example, by evaporating the mix placed in an open pan. Interpretation is performed as described in Chapter 1, 'Constant injection rate', or Chapter 2, 'Tracer gas dilution'. The problem is that the injection rate, I, is rather difficult to control for each component in the mix.
2   Pulse injection – sampling is started first. Soon after the sampling starts, a known mass of VOC mix is injected, either with a spray or, better, by flash evaporation (see Figure 6.11). This injection does not need to be very short, but it should be shorter than the sampling time. Sampling is continued after the end of the injection for more than two air changes, ensuring that VOC concentration is returned down to background level. Interpretation is performed as described in Chapter 1, 'Pulse injection'.

The amount injected is such that the resulting concentration values are clearly above the concentration outdoors, but below the saturation limit of the samplers.

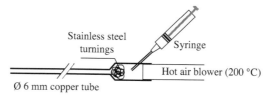

Figure 6.11 *Flash evaporation device for injecting the VOCs*
Source: Roulet *et al.*, 2000.

Air sampling and analysis

Air at the four locations is sampled with a pump through small tubes filled with an adsorbing medium (for example, activated charcoal, TENAX). The sampling rate is about 0.11 of air per minute. VOCs accumulate in the compound by absorption, as long as the medium is not saturated.

The sampling tube is then hermetically sealed and taken to the laboratory for further analysis. The VOCs are desorbed by heating the tubes and they are stored in a cold trap. The content of the trap is then injected in the column of a gas chromatograph. A flame ionization detector (FID) is used to detect and measure the amount of each compound, while a mass spectrograph is used to help identify the compound (Mogl *et al.*, 1995). An FID analyser can also be used on site to sample and analyse the air but, if pulse injection is used, this analysis should be performed at periodical and short intervals at the four locations shown in Figure 6.10.

Which VOCs?

The 'natural' concentration of VOCs is often (and hopefully!) not large enough to provide accurate measurement of transfer rate. Therefore, a mix of various VOCs should be injected in the extract duct in order to obtain concentrations that are significantly larger than concentrations in outdoor air. Criteria to determine the VOCs' mixture are:

- The compounds are selected on their occurrence in buildings. Sources are paints, paper, solvents, carpets and human emissions.
- They are representative of the different organic families with the characteristic of their functional group and saturation degree. The result is different boiling point and polarity.
- They are easy to analyse and the results are significant. The concentrations chosen must be under the saturation limit of the TENAX sampler.
- The compounds should be easy to manipulate
- They should not give a long-lasting and bad smell in the room and should present an acceptable toxicity at the measured concentrations.

Lists of VOCs found most often in office buildings, as well as proposals for VOC cocktails can be found in the literature (Brown *et al.*, 1994; Hodgson, 1995; Maroni *et al.*, 1995; Molhave *et al.*, 1997; Wolkoff *et al.*, 1997; van der

Table 6.2 *List of VOCs used for contaminant transfer experiments*

Class	Compound (current names)	Formula	Sources	Boiling point [°C]
Alkanes	n-decane	$C_{10}H_{22}$	Paints and associated supplies	174
Alcohols	n-butanol	$CH_3(CH_2)_3OH$	Adhesives	117
	1-hexanol	$CH_3(CH_2)_5OH$	Solvent	158
	phenol	C_6H_5OH		182
Haloalcanes	1,6-dichlorohexane	$Cl(CH_2)_6Cl$	Solvent, adhesives	203
Aldehydes	hexanal (caproaldehyde)	$CH_3(CH_2)_4CHO$	Paper, paints	128
	benzaldehyde	C_6H_5CHO	Adhesives	179
Cyclic and aromatic	4-isopropenyl 1-methyl cyclohexene (limonene)	$C_{10}H_{16}$	Perfumed waxes and cleaners	177
hydrocarbons	1,3-dimethylbenzene (m-xylene)	$1,3\text{-}(CH_3)_2C_6H_4$	Solvent, fuel	139
	1,3,5-trimethylbenzene (mesitylene)	$1,3,5\text{-}(CH_3)_3C_6H_3$	Solvent, fuel	165
Ethers	dipropylether	$CH_3(CH_2)_2O(CH_2)_2CH_3$	Paints and associated supplies	91

Wal *et al.*, 1998). These lists can inspire a selection of compounds to include in the mix. However, the total number of compounds should not be too large in order to keep the analysis at a practical level. Therefore, the list in Table 6.2, which includes most significant VOCs for several classes of organic compounds, was found suitable after careful selection.

It should be noted that acetic and butyric acids were chosen in a first step, but eliminated after the initial experiments for two reasons. First, because they are poorly selected by the columns used in the gas chromatograph, and second, because butyric acid has an awful smell.

Example of application

Experiments were performed on three small to medium air handling units, ventilating an auditorium and a laboratory at the EPFL,[1] and a building at the EMPA.[2] The characteristics of these units, measured using tracer gas dilution technique, are summarized in Table 6.3.

Both EPFL units are equipped with a rotating heat exchanger, whose wheel has a diameter of 785 mm, is 200 mm thick and rotates at 11 rpm. It is made of thin aluminium sheet, sinusoidally corrugated with a wavelength of 4.2 mm and peak-to-peak amplitude of 1.9 mm. A flat foil is placed between corrugated foils (see Figure 6.12). The aluminium sheets are treated to have a hygroscopic surface (see Figure 6.13)

The purging sector is a rectangle 100 mm high and 40 mm thick. The purging sector was correctly used in the laboratory unit. The auditorium

Table 6.3 *Characteristics of the air handling units used for the experiments*

		EPFL		EMPA
Quantity	**Unit**	**Auditorium**	**Laboratory**	**Akademie**
Outdoor airflow rate	m³/h	1900 ± 100	2530 ± 80	7000 ± 1000
Supply airflow rate		2070 ± 70	2900 ± 200	9500 ± 120
Extract airflow rate		1790 ± 40	1860 ± 50	5790 ± 60
Exhaust airflow rate		1600 ± 200	1500 ± 200	3000 ± 1600
Recirculation flow rate		130 ± 50	100 ± 200	1500 ± 900
Recirculation rate		7% ± 4%	5% ± 11%	26% ± 16%
Room mean age of air	min	15 ± 1	10 ± 1	20 ± 1

unit, however, was found with the wheel turning in the wrong way, making the purging sector inactive.

The measured ventilation efficiency in the EMPA academy is about 70 per cent, showing displacement ventilation. The rotating heat exchanger is made of Hexcore, a honeycomb consisting of a synthetic fibre material (Nomex). It has no hygroscopic coating. The wheel has a diameter of 1580 mm, is 140 mm thick

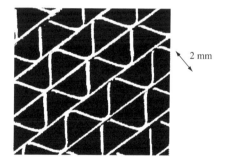

2 mm

Figure 6.12 *EPFL wheel structure*

10 μm

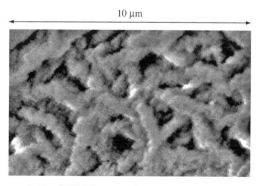

Figure 6.13 *SEM image of a new hygroscopic coating*

Table 6.4 *Pressure differentials in the units [Pa]*

	Filters	Auditorium unit in	Auditorium unit out	Laboratory unit in	Laboratory unit out	EMPA unit in	EMPA unit out
Across wheel	Inlet – supply	88 ± 5	80 ± 5	85 ± 5	82 ± 5	97 ± 2	109 ± 3
	Extract – exhaust	107 ± 5	110 ± 5	60 ± 5	67 ± 5	94 ± 1	88 ± 1
Between supply and exhaust	Cold side	125 ± 5	54 ± 5	30 ± 5	21 ± 5	-230 ± 2	-283 ± 5
	Warm side	-72 ± 5	-125 ± 5	-125 ± 5	-137 ± 5	-423 ± 2	-475 ± 5

and rotates at 5 rpm. The diameter of one honeycomb cell is 1.5 mm. The wheel has no purging sector and the fans are not in their ideal positions. This explains the high recirculation flow rate of about 15 per cent.

Pressure differentials were measured in all units just before or after the experiments. These are shown in Table 6.4. It can be seen that a negative pressure between supply and exhaust ducts allows vitiated air to pass from exhaust to supply through possible leakage. This is especially remarkable in the EMPA unit, where the fans are both on the same side of the wheel.

The climatic conditions just before or after the experiments are shown in Table 7.2.

Results

Experiments were performed several times in various conditions. In order to avoid adsorption in filters, these were taken out during experiments. Experiments were performed with the rotating heat exchanger turning in the correct direction, i.e. with an active purging sector, and in the wrong direction, thus suppressing the effect of the purging sector.

Table 6.5 *Climatic conditions in the units [°C]*

Unit	Auditorium unit					Laboratory unit				EMPA
Expt.	A + 1	A + 2	A + 3	A − 1	A − 2	L + 1	L + 2	L − 1	L − 2	
Dew point										
Inlet	—	—	0.1	—	0.1	3.8	−2.5	3.8	−2.5	1.8
Supply	—	—	1.1	—	1.1	3.8	−0.9	3.8	−0.9	11.7
Extract	—	—	2.3	—	2.3	4.6	−0.2	4.6	−0.2	11.2
Exhaust	—	—	0.9	—	0.9	3.8	−1.6	3.8	−1.6	12.1
Temperature										
Inlet	4.4	4.4	9.7	4.4	9.7	15.2	13.1	15.2	13.1	19.5
Supply	20.0	20.0	20.4	20.0	20.4	21.0	22.2	21.0	22.2	24.0
Extract	18.0	18.0	20.2	18.0	20.2	22.4	22.2	22.4	22.2	24.6
Exhaust	10.0	10.0	13.0	10.0	13.0	18.1	15.5	18.1	15.5	20.7

Table 6.6 *VOC transfer rate in the experiments performed in both EPFL units (%)*

| Purging sector | Auditorium unit | | | | | Laboratory unit | | | |
| | With | | | Without | | With | | Without | |
Experiment no.	A + 1	A + 2	A + 3	A − 1	A − 2	L + 1	L + 2	L − 1	L − 2
1,6-Dichlorohexane	26	16	38	−4	58	8	17	34	47
1-Butanol	9	10	29	16	42	11	9	34	39
1-Hexanal	15	14	23	15	29	17	3	38	34
1-Hexanol	36	32	45	30	54	16	20	41	62
Benzaldehyde	10	12	31	18	49	8	10	28	39
Dipropylether	10	13	13	5	22	5	5	24	9
Limonene	11	10	13	16	14	3	0	14	13
Mesitylene	5	9	20	15	31	7	9	28	28
m-Xylene	4	8	13	10	20	9	9	28	27
n-Decane	8	10	25	16	42	8	10	27	29
Phenol	51	44	54	50	71	32	37	64	64

Results of all experiments are shown in Tables 6.6 and 6.7.

Evidence for adsorption

Leakage and entrained air would result in the same recirculation rate for all chemical compounds, this rate being about 7 per cent for the auditorium unit, close to zero for the laboratory unit, and about 25 per cent for the EMPA unit. Recirculation rates of most VOCs are larger than that.

The differences among compounds shown in Figures 6.14 and 6.15 can be explained only by a physico-chemical behaviour such as adsorption. In all

Table 6.7 *VOC transfer rate in the EMPA experiments*

Experiment no.	1 (%)	2 (%)	3 (%)	Average (%)	Standard Deviation
Sampling time	91 mins	91 mins	45 mins		
1-Hexanal	74	57	92	74	18
1-Hexanol + m-Xylol	51	39	59	49	10
Benzaldehyde	85	63	100	83	19
Dipropylether	−15	25	53	39	20
Limonene	35	22	47	35	13
Mesitylene	46	29	54	43	13
n-Decane	46	30	54	43	12
Phenol	94	57	101	84	24

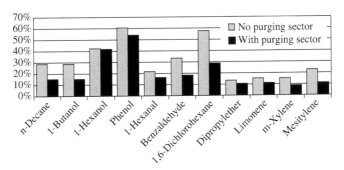

Figure 6.14 *Average VOC recirculation rates measured in the EPFL auditorium (leaky) unit, with and without purging sector*
Source: Roulet *et al.*, 2000.

experiments, the smallest recirculation rates are for limonene. They are close to the rates that could be expected from leakage or entrained air.

At the other extreme, it is clear that phenol, hexanal and dichlorohexane present a strong adsorption, which cannot be completely removed in the purging sector.

More evidence for adsorption is the dependence on boiling point shown in Figure 6.16. The transfer ratio for each type of compound increases with the boiling point of the compound, as can be expected for adsorption or condensation.

It is often claimed that non-hygroscopic wheels have a lower transfer ratio than hygroscopic wheels. Alternatively, it is also said among rotating heat exchanger specialists that non-hygroscopic wheels become hygroscopic with time (Ruud and Carlsson, 1996). It should be noticed that the recirculation rates measured in similar conditions (without purging sector) in a unit with a hygroscopic wheel (EPFL laboratory) and with a non-hygroscopic wheel (EMPA unit) are clearly correlated (see Figure 6.17). The line in this figure is a least square fit line, with a correlation coefficient of $R = 0.79$, and slope

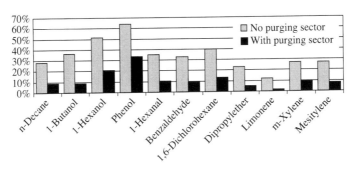

Figure 6.15 *Average VOC recirculation rates measured in the EPFL laboratory unit, with and without purging sector*

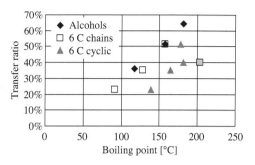

Figure 6.16 *Transfer ratio as a function of the boiling point for three families*

Note: The filled square corresponds to dichlorohexane.

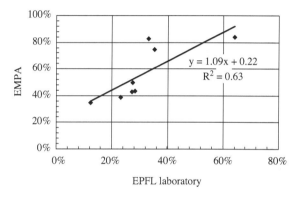

Figure 6.17 *Recirculation rates for each chemical compound measured in EPFL and EMPA units, in both cases without a purging sector*

unity. The average difference between recirculation rates is 22 per cent, close to the recirculation rate resulting from internal leakage in the EMPA unit.

This supports the hypothesis that the surface of the clean new wheel has not much influence on the adsorption properties of aged wheels. There are, however, new developments of rotating heat exchangers transferring latent heat and water vapour, without transferring chemical compounds such as odours (Okano *et al.*, 1999; Seibu Giken Co., 1999).

Strategies to improve the performance of HVAC systems

Indoor air quality strategies for an optimal performance of HVAC systems and their components can be divided in two categories:

- Strategies that affect the HVAC system while in use – operation, maintenance and replacement strategies;

Table 6.8 *General IAQ strategies for HVAC systems*

Design	Operation
Prevent pollution from outdoor air coming into the system	
Select appropriate filtering system	Discontinuous sources: ventilate mainly
Locate outdoor air intake at a clean site,	when the source intensity is small
far from potential pollution sources	Continuous sources: use an appropriate
	filtering system
Prevent pollution from recirculation	
Avoid recirculation	Suppress recirculation if possible
Install appropriate filtering system in	Discontinuous indoor air sources: no
extract duct. Active charcoal absorbs	recirculation at certain hours
odours	Continuous indoor air sources: use an
Install possibility for switching of	appropriate filtering system
recirculation system at certain hours	
System settings/operation strategies	
Room temperature 20°C (lower air temp.	Start system before official business hours
improves perceived air quality)	to purge the building (for example, two
Room humidity at 30–60% (lower room	nominal time constants earlier)
humidity improves perceived air quality,	Switch off during certain periods during
but lower than 30 per cent might affect	the day, when nobody is present (meeting
health negatively).	rooms, for example)

- Strategies that affect the design of the HVAC system – design principles and innovative design strategies.

For both of these categories, IAQ strategies were defined to prevent HVAC systems and components from being and becoming a source of pollution (Bluyssen *et al.*, 2003). These are summarized in Tables 6.8 to 6.14.

Table 6.9 *Checkpoints in HVAC units in visual inspection*

Component	passed	fail
Outdoor chamber		
Access door: exists but closed	☐	☐
Chamber is clean (no dirt, leaves, loose insulation materials etc.)	☐	☐
Appropriate water drainage	☐	☐
Filters		
Access door: closed airtight	☐	☐
Pressure difference meter is installed and operational	☐	☐
Chamber is clean and dry (no water on the bottom of the chamber)	☐	☐
No broken filters or visible leaks	☐	☐
Filters bags do not touch on the bottom of the chamber when the unit is not in operation	☐	☐
Filters are clean (commissioning) or not too dirty/smelly (100% of the back side is not dark)	☐	☐
Filters are of class F7 or better (fine filters)	☐	☐
Coils		
Access door: exists, but closed airtight	☐	☐
Fins are clean and not damaged	☐	☐
No visible microbial growth (cooling coils)	☐	☐
Drip pan and drainage (cooling coils)	☐	☐
Rotating heat exchanger		
Access doors on both sides	☐	☐
No significant leakage between supply and exhaust channels	☐	☐
Purging sector is installed on the warm side of the wheel	☐	☐
Wheel rotates in correct direction (wheel material passes in front of the purging sector after having left the exhaust part of the unit)	☐	☐
F5 or better class filters are installed upwind of the wheel in both ducts	☐	☐
Pressure difference between supply and exhaust is positive	☐	☐
Wheel stops automatically when heat recovery is smaller than the power needed to turn the wheel	☐	☐
Humidifier		
Access door exists, but closed airtight	☐	☐
Clean water, no visible microbial growth or oil film	☐	☐
Drop separator and drip pan are clean (if applicable)	☐	☐
Drainage with siphon; clean	☐	☐
Humidifier automatically shuts down if the air handling unit is off	☐	☐
The humidifier runs dry when the HVAC unit is shut down	☐	☐
Ducts		
Access door not blocked	☐	☐
No condensed water or microbial growth in the ducts	☐	☐
No pieces of construction materials, mineral wool etc in the ducts	☐	☐
No metal powder in the ducts	☐	☐
Amount of dust under limit value	☐	☐

Table 6.10 *IAQ strategies for filters*

Design	Operation
Prevent filters from becoming a source of pollution	
Select a low-polluting new filter	Change filter on time: depending on the
Condition (bake) new filters before use	situation, traffic and other loads once in
Use another filtering method such as	1–12 months, but in general every six
electrostatic filtering or apply a two or	months for highly polluted areas (town)
more phase filtering system	and one year for low-polluting areas
Keep the filter dry (by proper layout of air	(countryside)
intake section, heating of supply air with	Check pollution effect regularly in
pre-heater or heating filters for a certain	sensory, chemical and biological terms
period)	
Minimize micro-organisms (UV radiation)	
Avoid snow penetrating in the HVAC	
system by proper layout of air intake	
section	
Keep filter bags from lying on the bottom	
of the filter chamber when the HVAC unit	
is not in operation (preventing filters	
absorbing water due to eventual rain or	
snow penetration)	
Prevent outdoor air from bypassing the filter	
Make certain the filter frame and sealing	Make certain the filter frame and sealing
seat have no leaks	seat have no leaks
Prevent outdoor air impurities from	
passing the filter itself: proper choice of	
filter type	

Table 6.11 *IAQ strategies for ducts*

Design	Operation
Prevent ducts from becoming a source of pollution	
Use duct processed without oil, and which does not emit pollutants itself (label or smell)	Inspect ducts at least once a year and clean if necessary
Interior surfaces should be smooth, avoid sharp edges or self-tapping screws in ducts	
Prevent dust accumulating during operation or debris from construction	
Keep duct ends closed until in operation	Inspect ducts at least once a year and clean if necessary
Keep accessories packed in closed boxes	Check filter system that provides clean air to duct at least once a year
Remove packaging just before installation	
Prior to first operation, check all parts in contact with the airflow for cleanliness, and reclean if necessary	
Install a filter system upwind of the ducts	
Prevent condensation points	
Add insulation material outside the ducts	
Prevent condensation from humidifiers	
Other recommendations	
Limit flexible air ducts (difficult to clean)	Check location and service openings, especially in spaces with suspended ceilings Very often service openings in ducts are useless because there is no opening in the suspended ceiling or there are cables under the service opening
Avoid sealant with high emission and do not attach tapes or tags	
Install service openings	
Install stiffeners and other fittings in such a way that deposits of dirt are prevented and cleaning can be carried out	

Table 6.12 *IAQ strategies for rotating heat exchangers*

Design	Operation
Select a wheel equipped with purging sector and install it with the purging sector on the warm side of the wheel	If pressure on supply side is negative compared to exhaust side then change pressure hierarchy
Supply and exhaust fans should be located and sized so that a positive pressure difference of about 200 Pa is achieved between supply and exhaust ducts at the wheel level	If the rotation of wheel is in the wrong direction then change to proper direction: the wheel should pass from exhaust to supply ducts in front of the purging sector
Avoid hygroscopic wheels when contamination is an important concern	Clean dirty wheels according to instructions of the manufacturer, with either compressed air, vacuum cleaner or pressurized water
Change wheels if they are warped	
Install filters in both channels upwind of the heat exchanger	Check that the wheel control stops the wheel when no heat can be recovered

Table 6.13 *IAQ strategies for humidifiers*

Design	Operation
Prevent pollution from water, water tanks and devices/agents to disinfect, demineralize and/or desalinate the water	
Remove oil residue before use of humidifier to prevent an oil film on water	If there is an oil film on water surface, drain water and clean humidifier immediately
Take care using disinfecting material	
Use UV as germicide	Change water every week
De-ionization cartridges to prepare soft water might emit VOCs	Clean tank (not possible for steam: keep them dry/empty when not in use)
Install demineralization device in disperser to keep the oscillator circuit board free from mineral precipitation for as long as possible	Clean humidifier regularly every six months (dry or wet)
	Desalinization must take place with an agent that does not smell
Use a control system with which	
Humidifier should automatically shut down when the HVAC system is off	Check operation of control system regularly
New water is added when the water temperature exceeds 20°C (spray nozzle and evaporative humidifiers, ultrasonic humidifiers)	
Other recommendations	
Do not use porous wet material	
Prevent condensation from steam humidifier	

Table 6.14 *IAQ strategies for coils*

Design	Operation
Prevent water reservoirs and material of coils from becoming a source of pollution	
Keep outlet of drain free at the lowest point of the drain pan (include angle)	Maintain on time
Water/condensation should not stay too long in reservoirs: change system design	Water collection reservoirs: remove water regularly, clean
Remove oil before installation	Check for visible growth of moulds on coil surface
Prevent corrosion by selecting the proper material	
Do not place any adsorbing material behind cooling coil	
Prevent water droplets from being produced	
Place a droplet catcher behind cooling coil	

Notes

1 École Polytechnique Fédérale de Lausanne (Swiss Federal Institue of Technology of Lausanne).
2 Swiss Federal Research Institute for Materials.

7

Common Methods
and Techniques

Expressing concentrations and flow rates

Coherent units

When using equations, such as Equation 2.7, to model ventilation systems, coherent units should be used to get the correct results. Some examples are given in Table 7.1, and Annex A gives conversion tables. If the analysers and tracer gas flowmeters do not provide coherent units, the measured data should be converted to coherent units before further interpretation.

Corrections for density changes

Note that Equation 2.7 is essentially a mass conservation equation, and is therefore exact only when mass flow rates and mass concentrations are used. However, volume concentration and volume flow rate can be used as long as the density of air does not change too much along its flow.

The perfect gas law can model air as well as the tracer gases at ambient temperature:

$$pV = nRT \qquad (7.1)$$

where:

p	is the pressure of the gas,
V	is the occupied volume,
n	is the number of moles of gas,
$R = 8.31396 \, \text{J/(mole} \cdot \text{K)}$	is the molar gas constant,
T	is the absolute temperature.

The density of a mixture of gases, ρ can then be calculated. Let M_i be the molar mass of the gas, $\sum_i$, that is the mass of $N_{av} = 6.02486 \cdot 10^{23}$ molecules. The density is obtained by multiplying Equation 7.1 by M and dividing it by V:

$$\rho = \frac{m}{V} = \frac{\bar{M}n}{V} = \frac{\bar{M}p}{RT} \qquad (7.2)$$

Table 7.1 *Examples of coherent units*

Airflow rate	Injection rate	Concentration
kg/s	kg/s	Mass concentration
m^3/h	m^3/h	Volume concentration
m^3/h	cm^3/h	ppm

with

$$\bar{M} = \frac{\sum_i M_i n_i}{\sum_i n_i} \tag{7.3}$$

being the average molar mass of the mixture. For dry air, $\bar{M} = 28.96\,\text{g/mole}$. The relative change in density is then:

$$\frac{\Delta \rho}{\rho} = \frac{\Delta \bar{M}}{\bar{M}} + \frac{\Delta p}{p} - \frac{\Delta T}{T} \tag{7.4}$$

As long as the tracer gas is present only in trace concentration, the temperature has the largest effect on the density.

Conversion formulae for concentration

The conversion between units of tracer (or contaminant) concentration requires the knowledge of the densities of tracer (or contaminant) and air, or of their molecular masses. Therefore, formulae are given below instead of tables. Concentration of a gas in a mixture can be expressed in several ways:

The *mass concentration*, C_m, is the ratio of the mass of the considered gas, x, and the mass of the mixture:

$$C_m = \frac{m_x}{\sum_i m_i} \tag{7.5}$$

The *molar concentration*, C_M, is the ratio of the number of molecules, or of moles, of the considered gas to the total number of molecules, or moles, in the mixture:

$$C_M = \frac{n_x}{\sum_i n_i} \tag{7.6}$$

The *volume concentration*, C_V, is the ratio of the volume of the considered gas at the considered temperature and pressure and the volume of the mixture at the same temperature and pressure. It is also the ratio of the partial pressure of the considered gas and the total pressure:

$$C_V = \frac{V_x}{V} = \frac{p_x}{p} \tag{7.7}$$

In a mixture, every gas, x, occupies the whole volume, V:

$$V = \frac{n_x RT}{p_x} \tag{7.8}$$

The relations between mass concentration, C_m, molar concentration, C_M, and volume concentration, C_v of component x, are then:

$$C_m = \frac{m_x}{\sum_i m_i} = \frac{M_x n_x}{\sum_i M_i n_i} = \frac{M_x}{\bar{M}} C_M = \frac{M_x}{\bar{M}} C_V \tag{7.9}$$

Molar and volume concentration are the same, since at a given pressure and temperature, one mole of gas always occupies the same volume.

Tracer gas dilution techniques

Tracer gas dilution techniques are among the most efficient to assess airflow patterns within buildings and air handling systems. They consist of 'colouring' or marking the air with a tracer gas, i.e. a gas that mixes well with the air and is easy to analyse in trace amounts. Non-toxic tracer gases may be useful to simulate the behaviour of contaminants having similar densities.

The concentration of tracer gas is analysed when or where the tracer is well mixed with the air. The evolution of the measured concentration depends on both the injection flow rate and the airflow rate that dilutes the tracer. Interpreting this concentration evolution provides airflow rates, age of the air, ventilation efficiency, leakage flow rates and so on.

Applications of these techniques are presented in Chapters 1, 2 and 3. The technique itself, i.e. the tracer gases, the injection techniques and the analysers are presented here.

Properties of tracer gases

A tracer gas used for airflow measurements in buildings should ideally have the following properties:

1　be easily analysable, preferably at low concentrations to reduce cost and side effects such as density changes or toxicity;
2　have low background concentration, allowing the use of low concentration in measurements;
3　be neither flammable nor explosive at practical concentrations, for obvious safety reasons;
4　be non-toxic at the concentration used, for obvious health reasons in inhabited buildings;
5　have a density close to the air density (i.e. a molecular weight close to 29 g/mole) to ensure easy mixing;
6　not be absorbed by furnishings, decompose or react with air or building components;
7　should be cheap in the quantity required for measurement.

Table 7.2 *Properties of the gases most frequently used as tracers*

Tracer name	Chemical formula	Molecular weight	Density/air @NTP	MAC* [ppm]	MDC† [-]	Analyser (besides MS)
Helium	He	4	0.14	–	$>6 \cdot 10^{-6}$	
Neon	Ne	20	0.69	–	$>20 \cdot 10^{-12}$	
Carbon dioxide	CO_2	44	1.53	5000	$3 \cdot 10^{-6}$	IR
Nitrous oxide	N_2O	44	1.53	25	$50 \cdot 10^{-9}$	IR
Sulphur hexafluoride	SF_6	146	5.10	1000	$0.1 \cdot 10^{-12}$	ECD (IR)
Freon R11	$CFCl_3$	137	4.74	1000	$1 \cdot 10^{-12}$	ECD (IR)
Freon R12	CF_2Cl_2	120	4.17	1000	$50 \cdot 10^{-9}$	ECD (IR)
Freon R13	CF_3Cl	104	3.60	1000	$50 \cdot 10^{-9}$	ECD (IR)
Freon R22	CHF_2Cl	86	2.99	1000	$20 \cdot 10^{-9}$	ECD (IR)
Freon R111	CCl_3-CCl_2F	220	7.60	1000		ECD (IR)
Freon R112	CCl_2F-CCl_2F	203	7.03	1000	$50 \cdot 10^{-9}$	ECD (IR)
Freon R113	$CCl_2F-CClF_2$	187	5.90	1000	$50 \cdot 10^{-9}$	ECD (IR)
Freon R114	$CClF_2CClF_2$	171	5.90	1000	$50 \cdot 10^{-9}$	ECD (IR)
Freon R115	$CClF_2CF_3$	154	5.31	1000		ECD (IR)
Halon 1211	CF_2BrCl	165	5.53	?	$0.5 \cdot 10^{-9}$	ECD (IR)
Halon 1301	CF_3Br	149	4.99	?	$10 \cdot 10^{-12}$	ECD (IR)
Perfreons			**Liquid @NTP**			
PB Perfreobenzene	C_6F_6	186	(6.4)			ECD
PMB Perfluoromethylbenzene	$CF_3C_6F_5$	236	(8.1)			ECD
PMCH Perfluoro-methyl-cyclohexane	$CF_3C_6F_{11}$	350	(12.1)	–	10^{-14}	ECD
PDCH Perfluoro-dimethyl-cyclohexane	$CF_3CF_3C_6F_{10}$	400	(13.8)	–	10^{-14}	ECD
PMCP Perfluoro-methyl-cyclopentane	$CF_3C_5F_9$	300	(10.3)	–	10^{-14}	ECD
PDCB Perfluoro-dimethyl-cyclobutane	$CF_3CF_3C_4F_6$	300	(10.3)	–	10^{-14}	ECD

Note: * MAC = maximum acceptable concentration for health safety; † MDC = minimum detectable concentration using the best available analyser. The useful concentration is about 100 times larger; IR: Infrared absorption spectrograph or photo-acoustic detector; ECD: Gas chromatography and electron capture detector.

Item 5 is important mainly if the concentration is relatively high (for example, 0.1 per cent or higher). For this reason and to achieve also items 3, 4 and 7, items 1 and 2 are essential.

Table 7.2 shows properties of tracers that have been used, together with appropriate detection methods. Note that the mass spectrometer (MS) can

Table 7.3 *Background concentration of some gases*

Gas	Formula	Rural concentration
Water vapour	H_2O	$20 \cdot 10^{-3}$
Argon	Ar	$9.3 \cdot 10^{-3}$
Carbon dioxide	CO_2	$350 \cdot 10^{-6}$
Helium	He	$5.24 \cdot 10^{-6}$
Methane	CH_4	$1.48 \cdot 10^{-6}$
Nitrous oxide	N_2O	$315 \cdot 10^{-9}$
Ozone	O_3	$35 \cdot 10^{-9}$
Nitrogen oxides	NO_x	$3 \cdot 10^{-9}$
Methyl chloride	CH_3Cl	$630 \cdot 10^{-12}$
Freon R12	CCl_2F_2	$305 \cdot 10^{-12}$
Freon R11	CCl_3F	$186 \cdot 10^{-12}$
Carbon tetrachloride	CCl_4	$135 \cdot 10^{-12}$
Chloroform	$CHCl_3$	$20 \cdot 10^{-12}$
Neon	Ne	$18 \cdot 10^{-12}$
Sulphur hexafluoride	SF_6	$850 \cdot 10^{-15}$
Halon 1301	CF_3Br	$750 \cdot 10^{-15}$
PDCH or Perfluoro-dimethyl-cyclohexane	$CF_3CF_3C_6F_{10}$	$22 \cdot 10^{-15}$
PMCH or Perfluoro-methyl-cyclohexane	$CF_3C_6F_{11}$	$4.5 \cdot 10^{-15}$
PMCP or Perfluoro-methyl-cyclopentane	$CF_3C_5F_9$	$3.2 \cdot 10^{-15}$
PDCB or Perfluoro-dimethyl-cyclobutane	$CF_3CF_3C_4F_6$	$0.34 \cdot 10^{-15}$

Source: Dietz *et al.*, 1983.

potentially analyse any tracer. Table 7.3 shows their background concentrations in outdoor air.

It can be seen in Tables 7.2 and 7.4 that no tracer complies with all the requirements. Moreover, because of possible interferences in the analyser used for multi-tracer experiments, the use of a specific tracer may forbid the use of several other interesting tracers.

A comparative experiment of the mixing of different tracers (SF_6, N_2O and He) was performed to study the effect of density (Niemelä *et al.*, 1990). This study shows that differences may occur when the tracer is at concentrations higher than 10 per cent, for example, where it is not well mixed with air at the injection location. However, density effect is not a major cause of error for tracer gas measurements, and mixing can be improved (see 'Mixing tracer gases', below).

Indeed, each tracer gas has some inconveniences: helium is too light and requires an expensive mass spectrometer for analysis; neon is expensive; CO_2 is very cheap to obtain and to analyse but has a large background; N_2O interferes with water vapour; SF_6 has a strong greenhouse effect; freons and halons destroy the ozone layer; and perfreons are adsorbed in furniture.

Table 7.4 *Qualities of some tracer gases*

Name	Compliance with the quality							
	No fire hazard	Low toxicity	Density close to air	No reactivity	Ease of use	Back-ground conc.	No local sources	Low cost
Helium	++	++	—	++	+	−	+	+
Neon	++	++	++	++	++	+	++	—
Carbon dioxide	++	−	+	−	++	—	—	++
Nitrous oxide	*	−	+	−	+	+	+	+
SF6	†	+	−	+	++	++	++	−
Freons R11, R12, R13	†	+	−	+	+	+	+	+
Freons R111 to R115	†	+	—	+	+	+	+	+
Halon BCF	†	+	—	++	+	++	+	−
Halon R13B1	†	+	—	++	+	++	+	+
Perfreons (PFT)	++	++	—	—	+	++	++	+

Note: ++ Very good for that property; + Good; − Not so good; — Poor; * Is not combustible but a strong oxidant at high concentration and temperature; † Is not combustible but decomposes in a flame, producing toxic chemicals.

Mixing tracer gases

Perfect mixing of tracer gas in the air of the measured zone or in the measured duct is essential when determining the airflow rates, but not for experiments to determine the age of the air. We found that in a closed, quiet, isothermal room, it may take several hours to mix a tracer gas into the air. In a 60 m³ room, this time is shortened to less than half an hour if a 100 W heat source (such as a quiet person or a light bulb) is present. A small 20 W fan like those used to ventilate the power supplies of computers reduces the mixing time down to five minutes.

Several methods can be used to improve or accelerate the mixing of tracer gases. The most widely used method is to inject the tracer upwind of a mixing fan, which can be a small 20 W cooling fan used in the electronic industry. Alternatives include portable oscillating fans. This method works perfectly but changes the thermal gradients in the measured zone, and may affect the air exchange rates. It should not be used during the measurement of the age of the air.

Mixing fans are not necessary if the injection nozzles are located at the locations where natural convection or mechanical ventilation provides signifi-cant air currents. Moreover, a continuous injection flow rate greatly assists the attainment of a uniform tracer gas concentration and is therefore preferred to pulse injection.

Quick mixing with the air around the injection port is obtained if the velocity of the tracer gas at the injection nozzle is large enough to create a turbulent jet (Silva and Afonso, 2004). For this purpose, the flow controlling valve and nozzle should be at the end of the injection tube, with the tube maintained under pressure. It is noted, however, that these two conditions complicate the experimental arrangement since control leads must extend to the end of the injection tube, and the system, under pressure, will be more sensitive to leaks.

In buildings with a large internal volume it may be necessary to discharge large amounts of tracer. If this is the case, then the following method may be used: the operator works out a zigzag or circular path through the area that will give good coverage of the building. The time taken to walk along the path is noted. The amount of gas required to dose the area is evaluated (from knowledge of the building volume and the required initial concentration), and the gas flow rate needed to discharge that volume of gas in the time taken to walk along the path is calculated. The gas cylinder is set to discharge the gas at the required rate and the operator walks along the path carrying the discharging cylinder. Some mixing of air and tracer will occur as a consequence of the movement of the operator through the building.

Mixing of tracer gas in ducts and air handling units is described in Chapter 2, 'Tracer gas injection ports'.

Tracer density problem

If the tracer gas density differs significantly from that of the air, that is, if its molecular weight differs much from 29g/mole, the concentrated tracer rises or falls (depending on its density) directly as it leaves the injection duct. Injection jets help to avoid this, while diffusers make the phenomenon worse, since they lower the injection speed of the concentrated tracer.

Since there is no non-toxic tracer gas with the density of air (apart from ethane, which is explosive, the closest are highly toxic carbon monoxide, hydrogen cyanide and nitrous oxide!), a simple approach is to dilute the tracer in air at about $1:10$ or more, and to use this diluted mixture. This not only adjusts the density, but also increases the injection flow rate – thus helping mixing – and makes the problem of flow control easier in relatively small rooms or for small concentrations.

In any case, the tracer concentration should not exceed a value that significantly changes the density of air. A proposed limit is:

$$C_{\text{lim}} \approx 3 \cdot 10^{-4} \frac{\rho_{\text{air}}}{\rho_{\text{tracer}}} \approx \frac{10^{-2}}{M_{\text{tracer}}} \tag{7.10}$$

where M is the molecular mass of the tracer gas, in grams per mole. The factor $3 \cdot 10^{-4}$ corresponds to a change in density for a temperature variation of 0.1 K in pure air. Such changes are very unlikely to have a significant effect on the airflows in a space. With common tracer gases, this limit concentration is much larger than the concentrations commonly used.

Several experiments have shown that if the tracer is properly injected, the errors caused by the tracer density are negligible in relation to other sources of error (Sandberg and Blomqvist, 1985).

Sampling methods

Samples of air containing tracer gases need to be taken for analysis. There are several sampling methods, each one being adapted to a particular purpose.

Grab sampling using hand pumps and bags is very cheap, easy to install and needs few materials in the field. This method can be used for decay measurements in no more than a few zones, and for constant emission provided conditions remain constant.

The passive sampling technique, which relies on adsorbing the tracers on a porous material, is used to sample the air continuously in such a way that the amount of tracer collected is proportional to the dose. An advantage of the passive (and also of active adsorbing) sampling is that, because of the storage in the adsorbing material, very tiny concentrations can be detected. The passive samplers and emitters are the only testing material and can be sent for analysis by mail.

The above methods are most suitable for a small number of measurements. For continuous monitoring of variable airflow rates in several zones over a long period of time, sampling networks using tubes and pumps are recommended. Such a sampling network is made of pipes returning from each zone to the analyser, and one or more pumps to draw the air–tracer mixture through these pipes.

Grab sampling

This technique does not require expensive equipment to be used on the measurement site. The tracer gas is initially injected into the space and allowed to mix with the air. Because this whole process is designed to be as simple as possible, rudimentary injection techniques are usually employed: releasing the tracer from a syringe, a plastic bag or a plastic bottle has shown itself to be adequate for the purpose.

When required, the air in the space is sampled using syringes, flexible bottles, air bags or chemical indicator tubes (see 'Chemical indicator tubes', below). The sample taken in this manner is intended to give an instantaneous picture of the tracer concentration at that time, hence the actual time taken to take the sample should be kept as short as possible.

After further defined periods of time, more samples can be taken. A minimum of two samples are required to evaluate the average air change rate between the sampling times, but often more are taken to ensure accuracy. The time interval between samples or the absolute time that samples are taken must also be recorded. Air samples are then returned to the laboratory for analysis.

Passive sampling

These sampling devices are metallic or glass tubes a few millimetres in diameter, partly filled with a given quantity of adsorbing material, such as activated charcoal. For transport and storage, these tubes are sealed with tight caps. Properly used passive samplers adsorb all the tracers that are in the air entering the sampler. They are used to obtain a quantity of tracer that is nearly proportional to the dose (that is, the time integral of the concentration) received during the measurement time.

Passive (or diffusive) sampling is initiated by opening one end of the tube for hours, days or weeks. Since the tracer reaching the adsorbent is adsorbed, there is a concentration gradient between the absorbent and the entrance of the tube. This leads to a diffusive flow of tracer, proportional to the concentration gradient.

Active sampling can be carried out with the same tubes by pumping the air through the tube. This technique is mainly used to achieve quick sampling. To ensure that the entire tracer contained in the air is trapped on the adsorbent, care must be taken not to sample too large a volume of air or not to pump the air through the adsorbent too fast, otherwise 'break through' will occur.

Networks, pumps and pipes

A star-type pipe network may be used in conjunction with valves and pumps to periodically collect samples of air in the monitored zones and to direct them to the analyser, one after the other.

Any small, airtight air pump is suitable to pump the sampled air to the analyser. Its model and size is chosen for a low working pressure, and with a flow large enough to flush the content of the pipe between two analysers. The working pressure is determined by the pressure drop through the sampling tubes and the analyser, and is usually less than 1000 Pa.

The tubing must be airtight and should not significantly adsorb or absorb the tracer gases. For these reasons, polyvinyl chloride (PVC) pipes should be avoided, as well as Teflon if freons or PFTs are used. Suitable materials are nylon and polyethylene. Metallic pipes can also be used, but they are more difficult to install. Never use tubes that have contained pure tracers, such as pipes once used for injection, since the small amounts of tracer gas absorbed in the plastic material will contaminate the sampled air. Such tubes should be marked or coloured and used exclusively for injection.

The inner diameter of these tubes may range from a few millimetres to 1 cm. Smaller pipes lead to larger pressure drops and need stronger pumps, whereas larger pipes need larger flows to flush the content of the pipe in a reasonable time. To avoid large pressure drops and noise, the average speed in the tubes should not exceed 5 m/s. The choice of tube size is influenced by the overall length of tubing required, i.e. by the size of the building. Some indications are given below.

For common tracer gases at temperatures commonly found in buildings, the pressure drop, Δp, for a length, L, of pipe of diameter, D, and volume flow rate, q, is given by:

$$\frac{\Delta p}{L} \cong 733 \cdot 10^{-6} \frac{q}{D^4} \tag{7.11}$$

As an example, for an airflow rate of 100 l/h, that is $28 \times 10^{-6}\,\mathrm{m^3/s}$, the minimum pipe inner diameter will be 2.6 mm to have an air speed of 5 m/s. In this case, the pressure drop will be 420 Pa/m, which may be too large in most buildings. A pipe with a 4 mm inner diameter will have an air speed of 2.2 m/s and a pressure drop of 80 Pa/m, which allows for 12 m long pipes with a pump allowing 1000 Pa under pressure at 10 l/h.

Injection and sampling port locations

To ensure the best possible mixing of tracer with air in the measured zone, the tracer gases should be injected at the locations where natural convection or mechanical ventilation provides significant air currents. Examples are ventilation inlets and the bottom of heating devices.

Sampling locations should be kept away from injection points, but at locations which are representative of the air in the zone or where mixing can be reasonably assumed to be good. Ventilation exhaust grilles are generally good locations.

If there is a convective loop in the measured room, it is convenient to place injection and sampling points on this loop but at two opposite points. However, the points should not be placed near a door or a window that can be opened during the measurement.

In a two-storey building with an open staircase, the upstairs tracer injection points should be placed close to the staircase, while the sampling points should be near the outside walls.

To obtain a more representative sample, or to inject at several locations in a zone, the sampling or injection pipe may be connected to a mixing box or manifold, from which several pipes, of the same length and diameter, go to various locations in the zone.

Injection and sampling sequence

Multi-zone tracer gas active measurement methods generally use only one analyser and often single bottles of each gas. This requires the zones to be scanned in sequence. There are several ways to plan these sequences, the two extremes being sequential or simultaneous operation (see Figure 7.1).

In sequential operation, one zone at a time, the sample tube for a given zone is pre-purged with a fresh sample prior to analysis. After this, some tracer gas is injected and the injection pipe is subsequently purged and the cycle repeated in another zone.

The simultaneous operation applies mainly to the constant concentration method. The air from zone i is pre-purged, while the air from the preceding zone, i-1, is analysed. At the same time, the amount of tracer to be injected to zone i-2 (already analysed) is calculated and delivered while the injection pipe of zone i-3 (already injected) is purged. This strategy is much more complex to control but is fast.

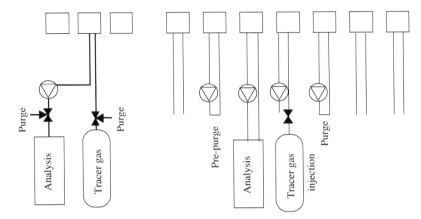

Figure 7.1 *Two strategies for injection and sampling*
Note: Left is one zone at a time; right is time shared.

Remember, however, that, in order to achieve good mixing, it is advantageous to inject the tracer continuously whenever possible. A series of short pulses spread evenly over time can simulate this continuous injection.

The air used to purge the pipes should ideally come from the measured zone and be returned to the same zone. However, if the flow rate in the sampling pipes is small, compared to the ventilation rate (which is the common case), it is reasonable to purge the injection tubes with outside air and to exhaust samples to the outside.

To ensure that the analysed air has a concentration averaged over the time between two measurements; it is possible to continuously pump the air from each zone into an inflatable bag that is then periodically emptied into the analyser.

Tracer gas analysers

Objective of the analysis

Measurement of the concentration of tracer in the air is the basic parameter needed for the interpretation or control of the concentration itself. Several analysing principles are available for such measurements. For any analyser, the salient features to consider are the following:

- Sensitivity – it is desirable to minimize the quantity of tracer gas used, not only from the point of view of cost, but also toxicity, fire or other hazards that may be relevant. The more sensitive the analyser, the lower is the required working concentration.
- Selectivity – the analyser should not be sensitive to other gases usually present in indoor air, for example, nitrogen, oxygen, water vapour, carbon dioxide, argon and so on.
- Speed – the time needed for the analysis must be considered, especially if several locations are to be sampled in sequence. The analysis time depends on the type of instrument and varies from milliseconds for mass spectrometers

up to several minutes for gas chromatographs or multi-tracer infrared analysers. Faster analysis will enable more frequent sampling of each zone and hence provide more detailed data. Frequent sampling (for example, every five to ten minutes) is essential for the constant concentration technique to maintain accurate control of concentration.

• Accuracy – last but not least, the accuracy of the concentration measurement directly influences the accuracy of the results.

There are several principles employed for analysis for tracer gas concentrations that differ in the gases analysed, the range of concentration detected, accuracy, speed, ease of use and cost. These principles are discussed and specific examples are given below.

Infrared absorption spectrometry

Any polyatomic gas molecule exhibits vibration modes, which are excited by infrared radiation. The wavelengths of the infrared radiation corresponding to molecular vibration frequencies[1] are absorbed in proportion to the number of tracer gas molecules present in the infrared beam. This absorption is used as a measure of the concentration of tracer gas molecules in the path between an infrared source and detector. This technique is referred to as absorption spectroscopy.

Infrared absorption spectrometers may be either dispersive or non-dispersive types, and both are in common use. Dispersive spectrometers use a diffraction grating that reflects the electromagnetic waves of a light beam into different directions, each direction corresponding to a given wavelength. The instrument is tuned by the operator to a narrow band of wavelengths specific to the gas of interest and any convenient infrared detector measures the absorption. Some modern instruments use tuneable laser diodes that emit the appropriate wavelength.

In non-dispersive devices, all the infrared radiation present in the absorption bands of the tracer is used. The infrared light beam is sent through both a reference channel containing nitrogen or pure air and an analysis channel that contains a sample of room air. A chopper is used to alternately pass the radiation from each channel to an analysis chamber containing a pure sample of the tracer gas. This gas heats and cools in response to the modulated beam. The heated gas expands through a measuring channel into an expansion chamber. The resulting alternating flow through the measuring channel is measured by a highly sensitive gas flow detector, which transmits an electric signal.

Characteristics of infrared absorption spectrometers:

• Analysable tracers – N_2O, SF_6, halon 1301, CO_2. Other detectable gases, such as H_2O, benzene, alkenes and so on, are not suitable as tracers.
• The sensitivity depends on the analysed gas. For example, full-scale deviation is 200 ppm for N_2O, or 20 ppm for SF_6 and CF_3Br.
• Interfering gases – care must be taken to eliminate the effects of other gases absorbing at similar frequencies (cross-sensitivity), particularly water

vapour and CO_2 present at high concentrations in the air. Filters are used to minimize the effect but humidity should be measured simultaneously to some tracers, such as N_2O, to allow for corrections.

- Analysis time – 10–50 s.
- Accuracy – ± 1 per cent of full scale if the zero drift is controlled.

Photo-acoustic detector

This analyser is also an infrared absorption spectrometer, but uses a different detector. An infrared radiation beam is first chopped then optically filtered to leave only frequencies that are absorbed by the tracer of interest. This beam then enters a gas-tight chamber containing the air sample. As above, the sample is heated and cooled in phase with the chopping frequency, creating sound waves in the chamber. This is the photo-acoustic effect. Microphones detect these sound waves.

Characteristics of analysers with photo-acoustic detector:

- Analysable tracers – N_2O, SF6, CO_2, freons F11, F112, 113 and 114, halons (one filter for each tracer). Other detectable gases are not suitable as tracers, such as H_2O, benzene, alkenes and so on.
- Sensitivity – the detection limit depends on the tracer but is typically 0.05 ppm, and the dynamic range is 10^5. The lowest full-scale range may then be 2 ppm but 10 ppm is recommended with usual tracers. The sensitivity for N_2O, CO and CO_2 drops strongly when these gases are diluted in dry nitrogen, as is often the case for calibration gases. Adding a special 'Nafion' tube in the sampling circuit allows for the moistening of the mixture in order to recover the normal high sensitivity.
- Interfering gases – several gases (which are not necessarily present in the air) may interfere with each tracer. Therefore, filters and tracers should be chosen in accordance with manufacturer's specifications.
- Analysis time – 30 s for one gas, 105 s for five gases and H_2O.
- Accuracy – ± 1 per cent of full scale.

Mass spectrometry

The pressure of the air sample is first lowered to about 10^{-5} Pa by pumping it through a capillary tube. The molecules of the sample are then ionized, accelerated to a given velocity and passed into a mass spectrometer.

The classical mass spectrometer curves the trajectory of the ions with a strong magnetic field. The radius of curvature depends on the velocity and charge-to-mass ratio of the ion; only those having the appropriate combination will pass a slit placed in front of the detector.

The most suitable spectrometer is, however, the quadrupole mass spectrometer, which is currently used in vacuum processes to analyse the residual gases. The gases entering into the analyser are ionized and the positive ions are separated by directing them axially between two pairs of rods creating an electric field at variable radio frequency. The ions follow a helicoidal path in

Table 7.5 *Tracer gases most used in the mass spectrometer technique*

Gas	Ions	Mass*	Comments
SF6	SF_5^+	127	
	SF^+	51	7.6% of mass 127 peak. Interferes with Freon 22
Freon R22	CHF^+	51	Interferes with SF_6
	$CHClF^+$	69	2.1% of peak 51. Interferes with R14 and R13B1
	$CHCl_2^+$	85	1.5% of peak at mass 51. Interferes with R12
Freon R12	$CClF_2^+$	85	Freon R13B1
	CF_3^+	69	
Freon R12B2$^+$	CF_2Br^+	120	Not commonly available
n-Butane	$C_4H_{10}^+$	58	Flammable above 2% concentration
	$C_3H_7^+$	43	
He	He^+	4	5.24 ppm background concentration
Ne	Ne^+	20	Expensive, $18 \cdot 10^{-12}$ background concentration
Ar	Ar^+	40	Background of 1%. Not a tracer but a good reference

Note: * Mass-to-charge ratio of the most common isotopes singly charged.
Source: Sherman and Dickerhoff, 1989.

this field. Only ions having a charge-to-mass ratio that corresponds to a given radio frequency reach an orifice at the end of that path and pass into an electron multiplier, whose signal is proportional to the number of incoming ions.

Such instruments deliver a signal for each molecule having a given charge-to-mass ratio. A given molecule may give several signals at different radio frequencies, since ionization may often break the molecule into several ions. For example, water vapour gives a peak not only at mass 18 (H_2O^+), but also at mass 17 (HO^+), 16 (O^+), 2 (H_2^+) and 1 (H^+). The electric current is proportional to the concentration, but the sensitivity depends on the analysed gas.

Characteristics of mass spectrometers:

- Analysable tracers – any tracer that can be distinguished from the normal components of air. Confusion may occur if the molecule or a part of it has the same charge-to-mass ratio as components of air. Examples are shown in Table 7.5.
- Sensitivity – 2×10^{-6} for tracers with low background concentration.
- Interfering gases – any gas present in the sample may interfere with another, but it is nevertheless possible to analyse up to seven tracers without there being too much interference from the gases in the air or between the tracers themselves.
- Analysis time – a few milliseconds.
- Accuracy – 1 per cent.

Gas chromatography

A puff of the sampled air is injected into a separating (chromatographic) column, a tube in which adsorbent material is packed. This column is heated and the pulse of sample is pushed with a flow of inert carrier gas. The various components of the sample pass through the column at various speeds according to their affinity for the adsorbent material. At the end of the column, the components emerge in sequence and can be quantitatively detected with a suitable detector. Both flame ionization detectors (FIDs) and electron capture detectors (ECDs) have been used for tracer gas analysis.

In the FID, a pair of polarized electrodes collects the ions produced when organic compounds are burned into an hydrogen flame, and the current produced is amplified before measurement. This detector is rugged, reliable, easy to maintain and operate, and is by far the most used in gas chromatography, since it combines a good sensitivity to organic compounds (limit of detection of about 10^{-9} g) with a good linearity within a range of up to 10^7.

However, organic compounds are not very good tracer gases and the ECD, which is much more sensitive to halogenated[2] compounds, is the most common in tracer gas analysis. In this detector, a radioactive nickel cathode emits electrons, which are received on an anode. Halogens capture these electrons, lowering the received current and thereby indicating the tracer concentrations. ECDs are popular since they can measure halogenated tracers to exceptionally low concentrations.

Characteristics of gas chromatographs:

- Analysable tracers (with ECD) – any halogenated compound like SF_6, freons, perfluorocarbons or perfluorocycloalkanes.
- Sensitivity – from ppb (10^{-9}) range for SF_6 down to 10^{-14} for the PFTs.
- Interfering gases – H_2O, O_2 (oxygen traps and desiccators are used to suppress these effects).
- Analysis time – a few minutes but can be lowered down to 20 seconds by shortening and back flushing the column, if high selectivity is not needed.
- Accuracy – depends on the quality of the calibration, but can be 2 per cent of reading.

Chemical indicator tubes

This is a single shot method to estimate the air change in a single zone by the decay or constant injection technique with some tracers.

Detector tubes are glass tubes packed with a selective solid absorbent, which gives a colour reaction to some gases. The tubes used are sensitive to CO_2 in the 0.01–0.30 per cent range. Tubes as supplied by the manufacturer are sealed at both ends. To make a measurement the seals are broken, one end of the tube (the correct end is indicated on the tube) is inserted into a pair of specially designed hand bellows, the other end being left open to sample the air tracer mixture.

By making the prescribed number of strokes of the hand-held bellows, the correct amount of air is drawn through the tube. This enables the tracer gas evaluation to be made. The glass tube has graduation marks on it, and the length of the discolouration caused by the reaction indicates the concentration of tracer in the room air. Detector tubes can only be used once and must be discarded after each sample taken.

This single shot method is not very accurate but it is cheap and easy to operate. Therefore, it is suitable for a rough first estimate of the air change rate. The interpretation of the result is performed using the integral decay method (see Chapter 1, 'Pulse injection').

Characteristics of chemical indicator tubes:

- Analysable tracers – CO_2, H_2O and many toxic gases that are not useable as tracers.
- Sensitivity – 0.01–0.3 per cent range for CO_2.
- Analysis time – one minute.
- Accuracy – ±5 per cent or 10 per cent of full scale.

Calibration of the analysers

Any analyser should be periodically calibrated by analysing standard samples, which are mixtures of the tracers in air or other inert gas. The calibration mixtures containing N_2O and CO_2 must be moistened when a photo-acoustic detector is used. To transfer the calibration mixture from the containers to the analyser, never use valves or tubes that were previously used with pure or high concentration tracers.

During the measurements, it is recommended to periodically sample and analyse the outside air as a convenient zero reference, even if no tracer is expected in the outdoor air.

Identification methods

Identification is to assess the values of some parameters in formulae from values of variables involved in these formulae. There are several identification techniques. Only some of them are presented below. The rationale of the presented methods can be found in the literature, so only the final formulae are presented here.

Linear least square fit

The problem is the following: given N pairs of data (x, y), find the straight line:

$$y = a + nx \tag{7.12}$$

fitting these points at the best. That means that the coefficients a and n should be such that the sum of the 'distances' of the measured points to the line is a minimum.

Linear least square fit of the first kind

Such methods are used to find the coefficients of leakage models of Equations 4.1 or 4.2 in fan pressurization (see Chapter 4, 'The fan pressurization method').

The regression of the first kind assumes that the abscissa, x_i, of each measurement is well known and that the distribution of the ordinates around the regression line is Gaussian with a constant standard deviation. This method is very commonly used but it should be emphasized that the above hypotheses are not verified in the case of permeability tests because the values of x_i are measured estimates.

The regression line of the first kind minimizes the sum of the square of the residual ordinates (vertical distances):

$$SSR_y = \sum_1^N [y_i - (a + nx_i)]^2 \tag{7.13}$$

Its coefficients can be calculated using the following relationships. First compute the estimates of the averages:

$$\bar{x} = \frac{1}{N} \sum_{i=1}^N x_i$$
$$\bar{y} = \frac{1}{N} \sum_{i=1}^N y_i \tag{7.14}$$

and the estimates of the variances:

$$s_x^2 = \frac{1}{N-1} \sum_{i=1}^N (x_i - \bar{x})^2$$

$$s_y^2 = \frac{1}{N-1} \sum_{i=1}^N (y_i - \bar{y})^2 \tag{7.15}$$

$$s_{xy} = \frac{1}{N-1} \sum_{i=1}^N (x_i - \bar{x})(y_i - \bar{y})$$

Then the best estimates of the coefficients a and n, according the above hypotheses, are:

$$n = \frac{s_{xy}}{s_x^2}$$
$$a = \bar{y} - n\bar{x} \tag{7.16}$$

The slope given by Equation 7.16 is valid if the x_i are exactly known, and the minimized distance is the sum of the square of the vertical distances between the measured points and the regression line.

Confidence in the coefficients

The variances on the linear coefficients of the regression of the first kind are usually estimated using the following relations, which assume that the dispersion around the line is Gaussian with a constant standard deviation and is the result of the measurement errors:

$$s_n = \frac{1}{s_x}\sqrt{\frac{s_y^2 - ns_{xy}}{N-2}} \quad \text{and} \quad s_a = s_n\sqrt{\frac{1}{N}\sum_{i=1}^{N} x_i^2} \tag{7.17}$$

If $T(P,\nu)$ is the significance limit of the two-sided Student distribution for a probability, P, for degree of freedom, ν, then the confidence levels on the coefficients are:

$$I_a = s_a T(P, N-2) \tag{7.18}$$

$$I_n = s_n T(P, N-2) \tag{7.19}$$

This means that with a probability, P, the coefficient, a, lies in the interval $[a - I_a, a + I_a]$ and the same for n.

The estimate of the variance around the regression line at abscissa x is:

$$s_y(x) = s_n\sqrt{\frac{(N-1)}{N}s_x^2 + (x - \bar{x})^2} \tag{7.20}$$

and the confidence interval in the estimate of y using the regression line for any x is:

$$I_y(x) = s_y(x)T(P, N-2) \tag{7.21}$$

The values of the two-sided Student distribution are given in Table 7.6. The relation (Equation 7.17) can be used to obtain the confidence intervals for a' and n' in the second kind regression if the roles of x and y are permuted.

Table 7.6 *Two-sided confidence limits $T(P, N-2)$ for a Student distribution*

	$T(P, N-2)$ for probability $P =$					
$N-2$	0.8	0.9	0.95	0.99	0.995	0.999
1	3.078	6.3138	12.706	63.657	127.32	636.619
2	1.886	2.9200	4.3027	9.9248	14.089	31.598
3	1.638	2.3534	3.1825	5.8409	7.4533	12.924
4	1.533	2.1318	2.7764	4.6041	5.5976	8.610
5	1.476	2.0150	2.5706	4.0321	4.7733	6.869
6	1.440	1.9432	2.4469	3.7074	4.3168	5.959
7	1.415	1.8946	2.3646	3.4995	4.0293	5.408
8	1.397	1.8595	2.3060	3.3554	3.8325	5.041
9	1.383	1.8331	2.2622	3.2498	3.6897	4.781
10	1.372	1.8125	2.2281	3.1693	3.5814	4.5787

Regression of the second kind

When there are uncertainties in both axes, there is no reason to emphasize the x axis, and the same procedure can be followed commuting the roles of x and y. Generally another regression line is obtained, which is given by:

$$y' = a' + n'x \tag{7.22}$$

with another pair of coefficients:

$$n' = \frac{s_y^2}{s_{xy}} \tag{7.23}$$

$$a' = \bar{y} - n'\bar{x}$$

This regression line minimizes the sum of the square of the residual abscissae:

$$SSR_x = \sum_{i=1}^{N} \left[x_i - \frac{y' - a'}{n'} \right]^2 \tag{7.24}$$

If the two lines are close to each other or, in other words, if:

$$n' \approx n \quad \text{and therefore} \quad a' \approx a \tag{7.25}$$

then it can be said that there is a good correlation between the two physical quantities x and y. The correlation coefficient defined by:

$$R = \frac{s_{xy}}{s_x s_y} \quad \text{hence} \quad R^2 = \left| \frac{n}{n'} \right| \tag{7.26}$$

is a measure of the interdependence of x and y. It is not a measure of the quality of the fit or of the accuracy of the estimates of the coefficients, since $|R| = 1$ for any fit based on only two sets of points. The estimates of the errors on a and n are calculated in 'Confidence in the coefficients', above.

Orthogonal regression

If the two regressions of the second kind are calculated and different results are obtained, the problem is to choose the coefficients: which pair is the closest to the reality? Since each pair of coefficients is obtained assuming that one variable is exactly known, it is likely that the best set is neither of them and instead lies in between, but where?

There are several answers to that question, none of them being really satisfactory. One recipe is to take an average slope:

$$\bar{n} = \frac{n + n'}{2} \tag{7.27}$$

or a weighted average slope:

$$\bar{n} = \frac{\varepsilon_y n + \varepsilon_x n'}{2} \tag{7.28}$$

where ε_y and ε_x are the experimental errors on y and x respectively and deduce a corresponding value of a using Equation 7.16. This recipe does not show clearly which quantity is minimized by the fit.

Another more physical way is so-called 'orthogonal' regression. It minimizes the real (orthogonal) distance between the measured points and the regression line drawn with the scales on the axes inversely proportional to the experimental errors, using variables weighted by the experimental errors:

$$Y_i = \frac{y_i}{\varepsilon_y} \quad \text{and} \quad X_i = \frac{x_i}{\varepsilon_x} \tag{7.29}$$

Writing the regression line with these coordinates:

$$Y = A + \nu X \tag{7.30}$$

the minimized residual is:

$$SSR_\perp = \sum_{i=1}^{N} \frac{(Y_i - \nu X_i - A)^2}{\nu^2 + 1} \tag{7.31}$$

The slope in original (x, y) coordinates is given by the following relations. Let us define:

$$\Delta n = \frac{s_y^2 - s_x^2}{2 s_{xy}} \tag{7.32}$$

If σ_x and σ_y are respectively the estimated standard deviations of the abscissas, x_i, and ordinates, y_i, the slope is given by:

$$n_\perp = \frac{\sigma_x}{\sigma_y} \left(\Delta n \pm \sqrt{\Delta n^2 + 1} \right) \tag{7.33}$$

Between the two possibilities, the sign of $n_\perp$ must be chosen equal to the sign of n given by Equation 7.16. This sign is always positive in pressurization tests. The coefficient, a, is obtained with the relation of Equation 7.16.

Bayesian identification

In section 'Linear least square fit', above, it was said that the usual regression techniques for the identification of the coefficients can strictly be used only when one of the variables is well controlled. In this case only, the relations given in 'Confidence in the coefficients', can be used to get a good estimate of the confidence intervals of the obtained coefficients.

When there are uncertainties on both axes, for example in pressurization tests, these methods are not strictly valid, since they do not give any information on the relation between the two coefficients and their uncertainties. If several measurements of the same leak are performed, several pairs of coefficients, C and n, for the relation:

$$Q = C \Delta p^n \tag{7.34}$$

will be obtained, and a correlation between C and n will be found: the larger C values correspond to the smaller n and vice versa. A good identification technique should give the most likely couple of coefficients together with the probability density $f(C, n)$. Such a technique exists (Tarantola, 1987) and is summarized below.

Identification of the model parameters

Let us put in a vector, $\mathbf{z}$, both measured data and model parameters that have to be determined and assume that this vector is a random variable with a normal distribution in k-fold space (k = number of parameters and data):

$$f(\mathbf{z}) = A \exp[-\tfrac{1}{2}(\mathbf{z} - \mathbf{z}_p)^T \mathbf{C}_z^{-1}(\mathbf{z} - \mathbf{z}_p)] \tag{7.35}$$

where:

$\mathbf{z}_p$ is the *a priori* vector, $\mathbf{z}$, containing the measured values and reasonable estimates of the parameters to be identified,

$\mathbf{C}_z$ is the covariance matrix between the elements of $\mathbf{z}$. Its diagonal elements are the variances of the measured quantities and *a priori* estimated variances of the parameters. These latter variances are generally large, since the parameters are generally not known before the measurements.

The components of the vector, $\mathbf{z}$, are linked by a mathematical model or a set of equations that can be written:

$$\rho(\mathbf{z}) = \mathbf{0} \tag{7.36}$$

For example, if a linear relationship is assumed between two measured variables, x and y, the set of equations:

$$y_i = a + bx_i \tag{7.37}$$

can be written in a matrix form:

$$\begin{pmatrix} 1 & x_1 & y_1 \\ 1 & x_2 & y_2 \\ \cdots & \cdots & \cdots \\ \cdots & \cdots & \cdots \\ 1 & x_n & y_n \end{pmatrix} \begin{pmatrix} a \\ b \\ -1 \end{pmatrix} = \mathbf{0} \tag{7.38}$$

Generally, the proposed model is not exact and it may be assumed that it has a normal distribution:

$$g(\mathbf{z}) = B \exp[-\tfrac{1}{2}\rho(\mathbf{z})^T \mathbf{C}_T^{-1} \rho(\mathbf{z})] \tag{7.39}$$

where $\mathbf{C}_T$ is the covariance matrix of the model. If the model is exact, this distribution is a Dirac distribution:

$$g(\mathbf{z}) = \delta[\rho(\mathbf{z})] \tag{7.40}$$

Combining the prior knowledge contained in the distribution $f(\mathbf{z})$ with the model described with the distribution $g(\mathbf{z})$ gives a new distribution containing the *a posteriori* information. This new distribution:

$$\sigma(\boldsymbol{z}) = C \exp\left[-\tfrac{1}{2}\{\boldsymbol{\rho}(\boldsymbol{z})^T \boldsymbol{C}_T^{-1} \boldsymbol{\rho}(\boldsymbol{z}) + (\boldsymbol{z} - \boldsymbol{z}_p)^T \boldsymbol{C}_z^{-1}(\boldsymbol{z} - \boldsymbol{z}_p)\}\right] \tag{7.41}$$

From this distribution, the $\boldsymbol{z}$ vector presenting the maximum likelihood can be found. It is the vector, $\boldsymbol{z}$, that minimizes the exponent:

$$\boldsymbol{\rho}(\boldsymbol{z})^T \boldsymbol{C}_T^{-1} \boldsymbol{\rho}(\boldsymbol{z}) + (\boldsymbol{z} - \boldsymbol{z}_p)^T \boldsymbol{C}_z^{-1}(\boldsymbol{z} - \boldsymbol{z}_p) \tag{7.42}$$

This most probable vector contains the identified model parameters and the most probable values of the measured quantities. Practically, this vector is found using numerical methods looking for the minimum of the exponent given above. More references on such methods are Mitchell and Kaplan (1969) and Nelder and Mead (1965).

Error analysis

This method allows us to obtain the *a posteriori* estimate, $\boldsymbol{C}_z^i$, of the co-variance matrix of the distribution, $\sigma(\boldsymbol{z})$. For that purpose, the model, $\boldsymbol{\rho}(\boldsymbol{z})$, is linearized around the most probable vector, $\boldsymbol{z}_s$. The *a posteriori* covariance matrix is then:

$$\boldsymbol{C}_z^i = (\boldsymbol{F}_s^T \boldsymbol{C}_T^{-1} \boldsymbol{F}_s + \boldsymbol{C}_z^{-1})^{-1} = \boldsymbol{C}_z - \boldsymbol{C}_z \boldsymbol{F}_s^T (\boldsymbol{F}_s \boldsymbol{C}_z \boldsymbol{F}_s^T + \boldsymbol{C}_T)^{-1} \boldsymbol{F}_s \boldsymbol{C}_z \tag{7.43}$$

where $\boldsymbol{F}_s$ is a matrix having the dimension $N \times M$, with $M = Nn + n + N$, N being the number of measurements and n the number of parameters to be identified. $\boldsymbol{F}_s$ contains the derivatives of the model, $\boldsymbol{\rho}(\boldsymbol{z})$, evaluated at the point, $\boldsymbol{z}_s$:

$$\boldsymbol{F}_s = \begin{pmatrix} \dfrac{\partial \rho_1}{\partial z_1} & \cdots & \dfrac{\partial \rho_1}{\partial z_M} \\ \cdots & \cdots & \cdots \\ \dfrac{\partial \rho_n}{\partial z_1} & \cdots & \dfrac{\partial \rho_N}{\partial z_M} \end{pmatrix}_{z_s} \tag{7.44}$$

Error analysis

Purpose of the error analysis

The accuracy of any measurement depends on the conditions in which the measurement is done, on the quality of the measuring instrument and on the skill of the people making the measurement. Measurements cannot be perfect, accuracy cannot be infinite, and any measurement result includes some uncertainty. That means that the result is not absolute, but it is always possible to state that the actual value is contained, with a given probability, within some confidence limits, or vice versa, that the probability that the actual value is outside some limits is lower than a certain value. Since this confidence interval may be large, there is no sense in giving the result of a measurement without any information on its reliability.

Generally, an instrument does not directly give the required information. In most cases, several measurements are combined to obtain the needed value. For example, in tracer gas measurements, several concentrations, gas flows, time and volume measurements are combined in equations that are solved to get the airflow rates. The errors accompanying the measured values propagate through the interpretation formulae and finally give a probable error on the final result.

In this chapter, some methods for estimating the error on the result are presented. Note that only the instrumental and random errors are treated here. Bias caused by misuse of the instruments or by a lack of precautions is not discussed here.

Definitions

Let x be the result of a measurement. If several measurements of the same physical quantity are made, the results, x_i, of these measurements will not be all equal, but nearly all of them will be within some interval. The *confidence interval* with a probability, P, has this probability to include the actual value. In practice, about NP results out of a large number, N, of measured values of the same quantity should be included in the confidence interval.

The confidence interval or the probable error can be expressed by two ways:

1   The *absolute error* is expressed in the same units as the physical quantity:

$$\text{Measurement} = x \pm \delta x \ [\text{unit}] \tag{7.45}$$

and the confidence interval goes from $x - \delta x$ to $x + \delta x$.

2   The *relative error* is the ratio of the absolute error to the measured value:

$$\varepsilon = \delta x / x \tag{7.46}$$

which can be expressed in per cent by multiplying ε by 100.

The inverse relation is:

$$\delta x = x\varepsilon \tag{7.47}$$

The results should always be given with their confidence interval (or with an estimate of the possible error) and with the unit used. The digits in the results should all be significant:

Correct: length $= 420 \pm 10\,\text{mm}$ or 420 mm within 2 per cent;

Not coherent: length $= 421.728 \pm 9.511\,\text{mm}$ or 421.728 mm

within 2.255 per cent.

A few statistics

Error analysis cannot be done well without using some basic statistical theory. There are simplified methods, which unfortunately often give too large an error

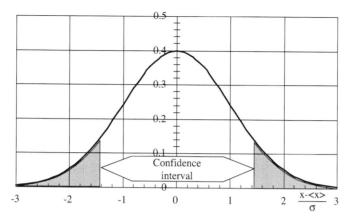

Figure 7.2 *Significance limits and confidence interval*

domain. The statistical method allows one to obtain more information on the reliability of the results.

Because of random reading errors and uncontrolled perturbations, the test values will follow a given distribution. We can model such distributions by treating x as a stochastic variable.

The *probability density function*, $f(X)$, of the variable x is the probability to find x between X and $X + dx$.

Its integral $F(X)$ is the probability of having $x < X$:

$$F(X) = \text{prob}(x < X) = \int_{-\infty}^{X} f(x)\, dx \tag{7.48}$$

The *lower significance limit* is the value X_i for which $F(X_i) = p$, where p is a given probability. The upper significance limit is the value X_s for which $F(X_s) = 1 - p$.

The *confidence interval* $[X_i, X_s]$ is the range between the lower and the upper significance limit (see Figure 7.2). The probability to find x in this interval is $P = 1 - 2p$.

Average

If the same importance is given to all the results, an estimate of the *average*, μ, of the variable, x, based on N measurements is calculated by:

$$\langle x \rangle = \frac{\sum_i x_i}{N} \cong \mu \tag{7.49}$$

where the sum runs over these N measurements ($i = 1, \ldots, N$).

If we give more importance to some measurements than to the others, a weight, w_i, can be attributed to each value, x_i, and the *weighted average* is calculated by:

$$\langle x \rangle = \frac{\sum_i w_i x_i}{\sum_i w_i} \tag{7.50}$$

Variance and standard deviation

A figure representing the importance of the scattering around the average value is the *mean square deviation* or *variance*:

$$S_x = \frac{\sum_i (x_i^2 - \langle x \rangle)^2}{(N-1)} = \frac{\sum_i (x_i^2) - N \langle x \rangle^2}{(N-1)} \tag{7.51}$$

The square root of S_x is the estimate, s_x, of the *standard deviation*, σ_x:

$$s_x = \sqrt{S_x} \cong \sigma_x \tag{7.52}$$

The larger the number of measurements, the better the estimate.

Covariance

An estimate s_{xy} of the covariance σ_{xy} of two random variables x and y, of which N measurements x_i and y_i were done, is calculated by:

$$S_x = \frac{\sum_i (x_i - \langle x \rangle)(y_i - \langle y \rangle)}{(N-1)} = \frac{\sum_i (x_i y_i) - N \langle x \rangle \langle y \rangle}{(N-1)} \tag{7.53}$$

This figure gives the tendency of two quantities to vary together. If these two variables are totally independent, the covariance will be zero. The covariance of a quantity with itself is the variance, already defined in Equation 7.51.

Statistical distributions

There are numerous probability distributions with a mathematical model. It is not the place here to present all of them. They can be found in the specialized handbooks such as Bevinton (1969), Diem and Lentner (1970) and Box *et al.* (1978). The two most used distributions, which are also used afterwards to estimate the confidence intervals, are presented below.

The probability density function of the *normal* or *Gaussian* distribution (see Figure 7.3, left) is:

$$f(c) = \frac{1}{\sigma \sqrt{2\pi}} \exp\left(-\frac{c^2}{2}\right) \qquad \text{where} \qquad c = \frac{x - \mu}{\sigma} \tag{7.54}$$

where μ is the average and σ the standard deviation of the variable x.

The probability of the normal distribution (see Figure 7.3, right) is:

$$F(c) = \frac{1}{2}\left[1 + \text{erf}\left(\frac{c}{\sqrt{2}}\right)\right] \tag{7.55}$$

where the error function erf(x) is:

$$\text{erf}(x) = \frac{2}{\sqrt{\pi}} \int_0^x \exp(-\xi^2) \, d\xi \tag{7.56}$$

with $\text{erf}(-x) = -\text{erf}(x)$.

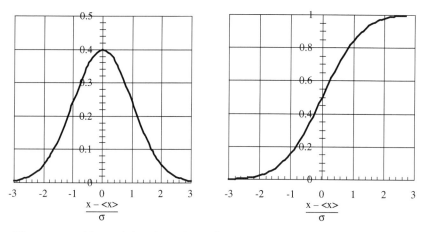

Figure 7.3 *Normal (or Gaussian) distribution (left) and its probabillity function (right)*

The confidence interval $[-c, c]$ of the normal distribution is obtained by solving the equation:

$$P = \text{erf}(c/\sqrt{2}) \tag{7.57}$$

for a given value of P.

If the normalized variable:

$$c = \frac{x - \mu}{\sigma} \tag{7.58}$$

is calculated using the estimate, s, of the standard deviation (based on $N + 1$ measurements) instead of the exact value, σ, (which is not known in practice), then this estimate of the normalized variable:

$$t = \frac{x - \mu}{s} \tag{7.59}$$

has a probability density function following the *Student distribution* (see Figure 7.4):

$$f(t, N) = \frac{\Gamma\left(\dfrac{N+1}{2}\right)}{\sqrt{N\pi}\,\Gamma\left(\dfrac{N}{2}\right)} \left(1 + \frac{t^2}{N}\right)^{N+1/2} t \tag{7.60}$$

where the gamma function $\Gamma(x/2)$ is:

if x is even: $\quad \Gamma(x/2) = (x/2 - 1)(x/2 - 2) \cdot \cdots \cdot 3 \cdot 2 \cdot 1$

if x is odd: $\quad \Gamma(x/2) = (x/2 - 1)(x/2 - 2) \cdot \cdots \cdot 1/2$ $\qquad (7.61)$

If n is large, the Student distribution tends to the normal distribution.

The confidence interval $[-T, T]$ where $T = T(P, \nu)$ of the Student distribution cannot be expressed analytically. It can be found in Table 7.6 (or in

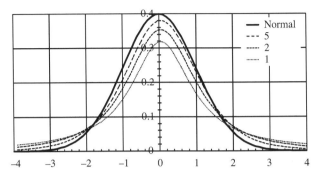

Figure 7.4 *Student distribution for 1, 2 and 5 degrees of freedom compared to the normal distribution*

more detail in statistical tables such as Zwillinger (2003)) and in most mathematical software packages.

Confidence interval of the Gaussian distribution

Assuming that a measurement, x_i, is a combination of the 'true' value μ and a random error, e_i, we have:

$$x_i = \mu + e_i \tag{7.62}$$

By measuring x_i, we expect to find the best estimate of μ. This can be done by performing $N > 1$ measurements and computing their average. This average x is the estimate of the 'true' value μ:

$$\mu \cong \langle x \rangle = \frac{\sum_i x_i}{N} \tag{7.63}$$

Let us recall that the confidence interval is the interval that has a given probability, P (for example, 95 per cent), to contain the 'true' value. In other terms, the probability of it being wrong, that is that the 'true' value being outside this confidence interval, is the error probability $p = 1 - P$.

What we need now is precisely to give the confidence interval around x that will contain μ with a fair probability. The value of this confidence interval depends on the probability distribution of the measured values. In principle, a reasonable distribution function should be chosen, adjusted on the measurements and the validity of this adjustment should be tested with the χ^2 test.

In most cases, however, and mainly when the number of the measurements is large, a normal distribution with a mean μ and a standard deviation σ can be assumed for the results of the measurements. Under this assumption, the confidence limit of the 'true' value is given by:

$$I_c = \frac{s}{\sqrt{N}} T(P, N - 1) \tag{7.64}$$

where s is the estimate of σ, and $T(P, N - 1)$ is the confidence interval of the Student distribution with $N - 1$ degrees of freedom. These are shown in Figure 7.5.

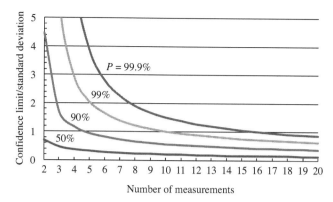

Figure 7.5 *Confidence limit divided by standard deviation versus number of measurements for various values of probability, P*

Hence, we can state:

$$\mu = \langle x \rangle \pm I_c \tag{7.65}$$

P is the probability that the confidence interval contains the 'true' value. P is chosen *a priori*, in practice between 0.9 and 0.99, depending on the degree of confidence needed. The higher the probability, the broader is the confidence interval $[-I_c, I_c]$.

Note that the confidence interval of the 'true' value stabilizes to a value close to the standard deviation if more than seven measurements are performed.

Error analysis

What is the problem?

If several measurements are combined to obtain the needed results, the errors should also be combined the proper way to get the resultant error. In other words, the problem is the following.

Suppose that we need several results $y_1, y_2, \ldots, y_j, \ldots, y_M$, each of them depending on measurements of several variables $x_1, x_2, \ldots, x_j, \ldots, x_N$:

$$y_j = f_j(x_1, x_2, \ldots, x_j, \ldots, x_N) \tag{7.66}$$

Here, j ($j = 1$ to M) enumerates the various results (for example, M different airflow rates) and i ($i = 1$ to N) enumerates the variables on which the results depend (for example, the tracer gas concentrations and flow rates or pressures and conductances).

If the measurements, x_i, each have an absolute error, δx_i, what are the errors, δy_j, on the results, y_j?

Most simple error analysis

The simplest rule, which is taught everywhere, is the following: the error δy on the result is estimated by replacing, in the total differential df of the function f, the infinitely small increments dx_i by the absolute error δx_i and by summing the absolute values:

$$\delta y_j = \sum_i \left| \frac{\partial f_j}{\partial x_i} \delta x_i \right| \tag{7.67}$$

If only arithmetical operations are used, the rules simplify to the following:

- If the result is obtained by adding or subtracting the measurements, the absolute error on the result, δy, is the sum of the absolute errors, δx, of each measurement.
- If the result is obtained by multiplying or dividing measured data, the relative error on the result, $\delta y/y$, is the sum of the relative errors, $\delta x/x$, on the measurements.

Estimate of the variance

The simplest method illustrated above is very rough, since it overestimates the confidence interval by supposing that all the errors in the measurements pull the result in the same direction, which is highly improbable. A statistical interpretation is then needed to take account of randomly distributed errors.

If the variances, s_{xi}, and the covariances, $s_{xi,xj}$, of the measurements are known or estimated, the covariances on the results, $s_{yk,yl}$, is, in a first approximation:

$$s_{yk,yl} = \sum_i \sum_j \frac{\partial f_k}{\partial x_i} \frac{\partial f_l}{\partial x_j} s_{xi,xj} \tag{7.68}$$

The variance of a given result is then:

$$(s_{yk})^2 = \sum_i \sum_j \frac{\partial f_k}{\partial x_i} \frac{\partial f_k}{\partial x_j} s_{xi,xj} \tag{7.69}$$

and if the measured variables are independent (that is if $s_{xi,xj} = 0$ when $i \neq j$), this simplifies to:

$$(s_{yk})^2 = \sum_i \left(\frac{\partial f_k}{\partial x_i} \right)^2 (s_{xi})^2 \tag{7.70}$$

The corresponding confidence intervals are then easily obtained by multiplying these results by the Student coefficient $T(P, \infty)$.

Linear equations systems

To interpret the results of an experience, we often should solve a system of equations such as:

$$Ay = x \tag{7.71}$$

where components of the vector x and the coefficients in the matrix A (which are the results of the measurements) are perturbed by random errors that can be represented by a vector δx and a matrix δA. The question is: which is the resulting error δy on the vector y, which is the vector containing the final results?

If the matrix δA and the vector δx were known, we could write:

$$(A + \delta A)(y + \delta y) = x + \delta x \tag{7.72}$$

and, taking Equation 7.71 into account, we could solve:

$$\delta y = (A + \delta A)^{-1}(\delta x - \delta Ay) \tag{7.73}$$

This equation can be used many times in a Monte-Carlo process, varying each time all the components of δA and δx at random, according to their probability density function. This provides a series of vectors δy from which an estimate of the probability density functions of the components can be calculated. However, this procedure is time consuming and, assuming a normal distribution of the measurement methods, simpler methods are available, which are described next.

Complete error analysis

The requested final result is calculated using:

$$y = A^{-1}x \qquad \text{hence} \qquad y_i = \sum_j \alpha_{ij} x_j \tag{7.74}$$

where the coefficients α_{ij} are those of the inverse matrix A^{-1}. The error calculated with the most simple (or the differential) method will then be:

$$\delta y_i = \sum_k \left| \frac{\partial y_i}{\partial x_k} \delta x_k \right| + \sum_{k,l} \left| \frac{\partial y_i}{\partial a_{kl}} \delta a_{kl} \right|$$

$$= \sum_k |\alpha_{ik} \delta x_k| + \sum_{k,l} \left| \frac{\partial \alpha_{ij}}{\partial a_{kl}} x_j \delta a_{kl} \right| \tag{7.75}$$

But since

$$\frac{\partial \alpha_{ij}}{\partial a_{kl}} = -\alpha_{ik} \alpha_{lj} \tag{7.76}$$

we get finally:

$$\delta y_i = \sum_k |\alpha_{ik} \delta x_k| + \sum_{kl} \left| \sum_j -\alpha_{ik} \alpha_{jl} x_j \delta_{kl} \right| \tag{7.77}$$

If the variances and covariances $s^2_{xk,xl}$, $s^2_{aki,amn}$ and s^2_{aki,x_m} are known, the covariance of the results $s^2_{yi,yj}$ is well estimated using a first order Taylor's expansion (Bevinton, 1969). We get:

$$s^2_{y_i,y_j} = \sum_{klmn} \frac{\partial y_i}{\partial a_{kl}} \frac{\partial y_j}{\partial a_{mn}} s^2_{a_{kl}a_{mn}} + \sum_{kl} \frac{\partial y_i}{\partial x_k} \frac{\partial y_j}{\partial x_j} s^2_{x_k x_l}$$

$$+ \sum_{klmn} \left(\frac{\partial y_i}{\partial a_{kl}} \frac{\partial y_j}{\partial x_m} + \frac{\partial y_j}{\partial a_{kl}} \frac{\partial y_i}{\partial x_m} \right) s^2_{a_{kl} x_m} \tag{7.78}$$

the partial derivatives are computed as above and we get finally:

$$s_{y_i,y_j}^2 = \sum_{klmn} \alpha_{ik} y_l \alpha_{jm} y_n s_{a_{kl} a_{mn}}^2 + \sum_{kl} \alpha_{ik} \alpha_{jl} s_{x_k x_l}^2$$
$$+ \sum_{klmn} (-\alpha_{ik} y_l \alpha_{jm} - \alpha_{jk} y_l \alpha_{im}) s_{a_{kl} x_m}^2 \tag{7.79}$$

which simplifies, if the variables are independent (that is, if the covariances are zero, which is not always the case):

$$s_{y_i,y_j}^2 = \sum_{kl} \alpha_{ik} \alpha_{jk} y_{l_n}^2 s_{a_{kl} a_{kl}}^2 + \sum_{kl} \alpha_{ik} \alpha_{jk} s_{x_k x_k}^2 \tag{7.80}$$

Upper bound of the errors

The vector δy contains a large number of data, but it is helpful to represent the error by a single value. To obtain such a single value, the following definitions, which can be found in the specific mathematical literature (for example, Deif, 1986) are used.

Vectorial norms and matrix norms

The norm $|x|$ of a vector x is any operation of the n-fold real space R^n in the ensemble of real numbers R satisfying:

$$|x| \geq 0 \text{ and } |x| = 0 \text{ if and only if } x = 0$$
$$|cx| = |c| \, |x| \text{ for any } c \in R \tag{7.81}$$
$$|x+y| \leq |x| + |y|$$

For example, the Euclidian norm that corresponds best to the standard deviation:

$$|x|_2 = \sqrt{\sum_i x_i^2} \tag{7.82}$$

complies with the relations in Equation 7.81, but there are many others, like $|x|_1 = \sum x_i$ or $|x|_\infty = \max(|x_i|)$.

The norm $|A|$ of a matrix A is any application $N(A) \Rightarrow |A| \in R$ satisfying:

$$|A| \geq 0 \text{ and } |A| = 0 \text{ if and only if } A = 0$$
$$|cA| = |c| \, |A| \text{ for any } c \in R$$
$$|A + B| \leq |A| + |B| \tag{7.83}$$
$$|A \cdot B| \leq |A| \cdot |B|$$

The matrix norm $|A|$ is *consistent* with the vectorial norm $|x|$ if:

$$|Ax| \leq |A| \cdot |x| \text{ for any } x \tag{7.84}$$

and the matrix norm is *subordinated* to the vectorial norm $|\boldsymbol{x}|$ if:

$$|\boldsymbol{A}| = \max \left(\frac{|\boldsymbol{Ax}|}{|\boldsymbol{x}|} \right) \text{ for any } x \neq 0 \tag{7.85}$$

The subordinated norm is the smallest matrix norm compatible with the norm $|x|$.

For example, the norm $|\boldsymbol{A}|_2$, defined as:

$$|\boldsymbol{A}|_2 = \sqrt{\lambda_1} \tag{7.86}$$

where λ_1 is the largest eigenvalue of $\boldsymbol{A}^H \boldsymbol{A}$ ($\boldsymbol{A}^H$ = hermitic conjugate or transpose of the complex conjugate matrix) is subordinated to the Euclidian norm $|x|_2$ but the Frobisher norm:

$$|\boldsymbol{A}|_F = \sqrt{\sum_i a_i^2 \sum_j a_j^2} \tag{7.87}$$

is consistent with the Euclidian norm but not subordinated to it.

Finally, the following norms, which lead to faster calculations, are often used:

$$|\boldsymbol{x}|_1 = \sum_i |x_i| \tag{7.88}$$

and the corresponding norm for the matrix $\boldsymbol{A}$:

$$|\boldsymbol{A}|_\infty = \max |\boldsymbol{A}_j| \tag{7.89}$$

where $\boldsymbol{A}_j$ are the column vectors of $\boldsymbol{A}$.

Calculation of the upper bound

From the norms of the experimental errors δy and $\delta \boldsymbol{A}$, it is possible to calculate an upper limit to the norm of the resulting error δx by the use of the following relation.

If $|\boldsymbol{I}| = 1$ (it is true for $|\boldsymbol{I}|_2$), then the norm of the relative error is:

$$\frac{|\delta x|}{|x|} \leq \frac{|\boldsymbol{A}| \cdot |\boldsymbol{A}^{-1}|}{1 - |\delta \boldsymbol{A}| \cdot |\boldsymbol{A}^{-1}|} \left(\frac{|\delta y|}{|y|} + \frac{|\delta \boldsymbol{A}|}{|\boldsymbol{A}|} \right) \tag{7.90}$$

The quantity:

$$\text{cond}(\boldsymbol{A}) = |\boldsymbol{A}| \cdot |\boldsymbol{A}^{-1}| \tag{7.91}$$

is of great importance in this calculation. It is the so-called *condition number* of the matrix $\boldsymbol{A}$ related to the used norm. This number indicates how nearly singular the matrix is.

If the spectral norm $|\boldsymbol{A}|_2$ is used, we get the smallest possible condition number, which is:

$$\text{cond}_2(\boldsymbol{A}) = |\boldsymbol{A}|_2 \cdot |\boldsymbol{A}^{-1}|_2 = \sqrt{\lambda_1 \lambda_n} \tag{7.92}$$

where λ_1 and λ_n are respectively the largest and the smallest eigenvalues of $A^H A$. This number is the *spectral condition number*.

Constant absolute error

If the absolute error is constant:

$$\delta A = e\mathbf{1} \quad \text{and} \quad \delta y = \delta y\mathbf{1} \tag{7.93}$$

where $\mathbf{1}$ and 1 are respectively a matrix of order N and an N-component vector with all elements equal to 1 (they are not the identity matrix and the unit vector).

It is easy to see that:

$$|\mathbf{1}|_2 = N \quad \text{and} \quad |1|_2 = \sqrt{N} \tag{7.94}$$

because the eigenvalues of $\mathbf{1}$ are N and 0^{N-1}, and those of $\mathbf{1}^H\mathbf{1} = \mathbf{1}^2$ are N^2 and 0^{N-1}. It follows that:

$$|\delta A|_2 = eN \quad \text{and} \quad |\delta y|_2 = \sqrt{N}\delta y \tag{7.95}$$

and

$$\frac{|\delta x|}{|x|} \le \frac{\text{cond}_2(A)}{1 - eN|A^{-1}|}\left(\frac{\sqrt{N}\delta y}{|y|} + \frac{Ne}{|A|}\right) \tag{7.96}$$

Constant relative error

In this case, if e and ε are the constant relative errors on A and y:

$$\delta A = eA \quad \text{and} \quad \delta y = \varepsilon y \tag{7.97}$$

and, from the definitions of the norms:

$$|\delta A| = e|A| \quad \text{and} \quad |\delta y| = \varepsilon y \tag{7.98}$$

and we get, for any norm satisfying $|I| = 1$:

$$\frac{|\delta x|}{|x|} \le \frac{\text{cond}(A)}{1 - \text{cond}(A)}(\varepsilon + e) \tag{7.99}$$

assuming that $e\,\text{cond}(A) < 1$, that is that $A + dA$ is regular.

Example: A measurement with two tracer gases at constant concentration in two zones gives the results in Table 7.7.

Table 7.7 *Data measured during a tracer gas experiment in two connected rooms*

	Zone 1	Zone 2
Tracer concentration 1	10 ppm	2.29 ppm
Tracer concentration 2	6.46 ppm	10 ppm
Injection rate, tracer 1	$2.65 \cdot 10^{-4}\,\text{m}^3/\text{h}$	$0\,\text{m}^3/\text{h}$
Injection rate, tracer 2	$0\,\text{m}^3/\text{h}$	$3.6 \cdot 10^{-4}\,\text{m}^3/\text{h}$

Table 7.8 *Airflow rates [m³/h], calculated from the data given in Table 7.7*

Flow going to	Outdoors	Zone 1	Zone 2
Flow coming from			
Outdoors	–	11.0	32.6
Zone 1	21.4	–	9.7
Zone 2	22.2	20.1	–

From which, using the method described in Chapter 1, 'Zone by zone systems of equations', we get the airflow rates to and from each zone [m³/h] shown in Table 7.8.

Let us suppose that the error on the injection rate is 5 per cent and if the relative error on the concentration is 2 per cent. What is the probable error on the airflows? Using the most simple method, we get a relative error of 9 per cent.

Using the spectral norm, we get relative errors of 12 per cent for zone 1 and 7 per cent for zone 2. These are upper limits that are easily calculated, but there is more information than the simple method allows, since we can see the difference in quality of the measurements between the two zones.

Notes

1 The wavelength, λ and the frequency, f, of an electromagnetic wave such as light, infrared or radio waves are related by $\lambda f = c$, where c is the velocity of light $(3 \cdot 10^{10}$ m/s).

2 Halogenated compounds are compounds containing fluorine, chlorine bromine and iodine in their molecule.

References

Aeschlimann, J.-M., C. Bonjour and E. Stocker, eds, 1986, *Méthodologie et Techniques de Plans d'Expériences: Cours de Perfectionnement de l'AVCP*, vol. 28, AVCP, Lausanne.

Andersson, B., K. Andersson, J. Sundell and P.-A. Zingmark, 1993, Mass transfer of contaminants in rotary enthalpy exchangers, *Indoor Air*, vol. 3, pp. 143–148.

ASHRAE, 2001, *Handbook – Fundamentals*, ASHRAE, Atlanta.

ASTM, 2003, *E1554-03 Standard Test Methods for Determining External Air Leakage of Air Distribution Systems by Fan Pressurization*, ASTM, Philadelphia.

Awbi, H., 2007, *Ventilation Systems, Design and Performance*, Taylor and Francis, London.

Axley, J. and A. Persily, 1988, Integral mass balance and pulse injection tracer technique, 'Effective Ventilation', 9th AIVC Conference, Gent, Belgium, AIVC.

Bandemer, H. and A. Bellmann, 1979, *Statistische Versuchsplanung*, BSB G. Teubner Verlag, Leipsig.

Bevinton, P., 1969, *Data Reduction and Error Analysis for the Physical Sciences*, MacGraw Hill, New York.

Björkroth, M., B. Müller, V. Küchen and P. M. Bluyssen, 2000, Pollution from ducts: What is the reason, how to measure it and how to prevent it?, *Healthy Buildings conference*, vol. 2, Espoo (SF), pp. 163–169.

Bluyssen, P. M., 1990, *Air Quality Evaluated by a Trained Panel*, Technical University of Denmark, Lund.

Bluyssen, P. M., E. De Oliveira Fernandes, L. Groes, G. Clausen, P.-O. Fanger, O. Valbjorn, C.-A. Bernhard and C.-A. Roulet, 1995, European audit study in 56 office buildings: Conclusions and recommendations, *Healthy Buildings conference*, vol. 3, Milano, pp. 287–292.

Bluyssen, P. M., C. Cox, J. Souto, B. Müller, G. Clausen and M. Bjørkroth, 2000a, Pollution from filters: What is the reason, how to measure and to prevent it?, *Healthy Buildings conference* vol. 2, Espoo (SF), p251.

Bluyssen, P. M., M. Björkroth, B. Müller, E. D. O. Fernandes and C.-A. Roulet, 2000b, Why, when and how do HVAC-systems pollute? Characterisation of HVAC systems related pollution, *Healthy Buildings 2000*, vol. 2, Espoo (SF), pp. 233–238.

Bluyssen, P. M., C. Cox, O. Seppänen, E. D. O. Fernandes, G. Clausen, B. Müller and C.-A. Roulet, 2003, Why, when and how do HVAC-systems pollute the indoor environment and what to do about it? The European AIRLESS project, *Building and Environment*, vol. 38, pp. 209–225.

Box, G. E. P., W. G. Hunter and J. S. Hunter, 1978, *Statistics for Experimenters: An Introduction to Design, Data Analysis and Model Building*, John Wiley, New York.

Brown, S. K., M. R. Sim, M. J. Abramson and C. N. Gray, 1994, Concentrations of volatile organic compounds in indoor air: A review, *Indoor Air*, vol. 4, pp. 123–134.

Carrie, F. R., P. Wouters, D. Ducarme, J. Andersson, J. C. Faysse, P. Chaffois, M. Kilberger and V. Patriarca, 1997, Impacts of air distribution system leakage in Europe: The SAVE Duct European Programme, *18th AIVC Conference* vol. 2, Athens, pp. 651–660.

CEN, 1999, *EN 832, Thermal Performance of Buildings: Calculation of Energy Use for Heating – Residential Buildings*, CEN, Brussels.

CEN, 2006, *prEN 15251, Criteria for the Indoor Environment, Including Thermal, Indoor Air Quality (Ventilation), Light and Noise*, CEN, Brussels.

CEN, 2007, prEN ISO/FDIS 13790, *Thermal Performance of Buildings: Calculation of Energy Use for Heating and Cooling*, CEN and ISO, Brussels and Geneva.

Davidson, L. and E. Olsson, 1987, Calculation of age and local purging flow rate in rooms, *Building and Environment*, vol. 22, pp. 111–127.

Deif, A., 1986, *Sensitivity Analysis in Linear Systems*, Springer Verlag, Berlin, Heidelberg and New York.

Dickerhoff, D. J., D. T. Grimsrud and R. D. Lipschutz, 1982, *Components Leakage Testing in Residential Buildings: Summer Study in Energy Efficient Buildings*, Lawrence Berkeley Lab, Santa Cruz, CA.

Diem, K. and C. Lentner, 1970, *Scientific Tables*, J. R. Geigy, Basel.

Dietz, R. N., R. W. Goodrich, and E. A. Cote, 1983, *Brookhaven Air Infiltration Measurement System (BNL/AIMS). Description and Application*, BNL report 33846, Brookhaven National Laboratory, Upton, New York.

Drost, M. K., 1993, Air to air heat exchanger performance, *Energy and Buildings*, vol. 19, pp. 215–220.

Elkhuizen, P. A., P. M. Bluyssen and L. Groes, 1995, A new approach to determine the performance of a trained sensory panel, *Healthy Buildings conference*, Milano, pp. 1365–1370.

Enai, M., C. Y. Shaw, J. T. Reardon and R. Magee, 1990, Multiple tracer gas technique for measuring interzonal air flows in buildings, *ASHRAE Trans.*, vol. 96, part 1, pp. 590–598.

Etheridge, D. and M. Sandberg, 1996, *Building Ventilation, Theory and Measurement Techniques*, J. Wiley & Sons, Hoboken, NJ.

Fanger, P. O., 1988, Introduction of the olf and decipol units to quantify air pollution perceived by human indoors and outdoors, *Energy and Buildings*, vol. 12, pp. 1–6.

Fedorov, V. V., 1972, *Theory of Optimal Experiments*, Academic Press, New York.

Feneuille, D., D. Mathieu and R. Phan-Tan-Luu, 1983, *Méthodologie de la Recherche Expérimentale*, Cours IPSOI, R. H. Poincaré, Marseille.

Fischer, T. and F. D. Heidt, 1997, Testing the ventilation efficiency of ventilation units with tracer gas methods, *Second International Conference, Buildings and the Environment*, Paris, pp. 405–413.

Frischtknecht, R., P. Hofstetter, I. Knoepfel, R. Dones and E. Zollinger, 1994, *Oekoinventare für Energiesysteme*, ETHZ, Zurich.

Fürbringer, J.-M. and C.-A. Roulet, 1991, Study of the errors occurring in measurement of leakage distribution in buildings by multifan pressurization, *Building and Environment*, vol. 26, pp. 111–120.

Fürbringer, J.-M., F. Foradini and C.-A. Roulet, 1994, Bayesian method for estimating air tightness coefficients from pressurisation measurements, *Building and Environment*, vol. 29, pp. 151–157.

Hakajiwa, S. and S. Togari, 1990, Simple test method of evaluating exterior tightness of tall office buildings, in E. M. H. Sherman, ed., *ASTM STP 1067*, ASTM, Philadelphia, pp. 231–248.

Hanlo, A. R., 1991, Use of tracer gas to determine leakage in domestic heat recovery units (HRV), in *Air Movement and Ventilation Control within Buildings*, AIVC 12th Conference proceedings, vol. 3, Ottawa, pp. 19–28.

Hodgson, A. T., 1995, A review and limited comparison of methods for measuring total volatile organic compounds in indoor air, *Indoor Air*, vol. 5, pp. 247–257.

ISO, 1977, *ISO 3966: Measurement of Fluid Flow in Closed Conduits, Velocity Area Method Using Pitot Static Tubes*, ISO, Geneva.

ISO, 1978, *ISO 4053: Measurement of Gas Flow in Conduits, Tracer Methods*, ISO, Geneva.

ISO, 1998, *ISO 9972: Thermal Insulation, Assessment of the Airtightness of Buildings, Fan Pressurisation Method*, ISO, Geneva.

ISO, 2003, *ISO 5167: Measurement of Fluid Flow by Means of Pressure Differential Devices Inserted in Circular Cross-section Conduits Running Full*, ISO, Geneva.

Jaboyedoff, P., C. A. Roulet, V. Dorer, A. Weber and A. Pfeiffer, 2004, Energy in air-handling units: Results of the AIRLESS European Project, *Energy and Buildings*, vol. 36, pp. 391–399.

Maroni, M., B. Seifert and T. Lindvall, 1995, *Indoor Air Quality: A Comprehensive Reference Book*, Air Quality Monographs, vol. 3, Elsevier, Amsterdam.

Mitchell, R. A. and J. L. Kaplan, 1969, Non linear constrained optimization by a non-random complex method, *J. Res. Natl. Bur. Stand.*, vol. C72, pp. 249–258.

Modera, M. P., 1989, Residential duct system leakage: Magnitude, impacts, and potential for reduction, *ASHRAE Trans.*, vol. 95, pt 2, pp. 561–569.

Mogl, S. K., J. Haas and R. Knutti, 1995, Bestimmung von VOC in Büroräumen (Analysis of VOC in Office Environment), *Mitt. Gebiete Lebensm. Hyg.*, vol. 86, pp. 667–671.

Molhave, L., G. Clausen, B. Berglund, J. D. Ceaurriz and A. Kettrup, 1997, Total volatile organic compounds in indoor air: Quality investigations, *Indoor Air*, vol. 7, pp. 225–240.

Müller, B., K. Fitzner and P. M. Bluyssen, 2000, Pollution from humidifiers: What is the reason; how to measure and to prevent it, *Healthy Buildings conference 2000*, vol. 2, Espoo (SF), p. 275.

NBCF, 1987, *Indoor Climate and Ventilation in Buildings: Regulations and Guidelines 1987*, The Finnish Ministry of the Environment, Helsinki.

Nelder, A. and R. Mead, 1965, A simplex method for function minimization, *J. Comput. J.*, vol. 7, pp. 308–313.

Niemelä, R., A. Lefevre, J.-P. Muller and G. Aubertin, 1990, Comparison of three tracer gases for determining ventilation effectiveness and capture efficiency, *Roomvent conference, Oslo*, 1990.

Okano, H., R. Kuramitsu and T. Hirose, 1999, New adsorptive total heat exchanger using exchange resin, *5th International Symposium on Separation Technology Between Korea and Japan*, Yonsei University Seoul, pp. 618–621.

Pejtersen, J., 1996, Sensory air pollution caused by rotary heat exchangers, *Indoor Air conference*, vol. 3, Nagoya, Japan, pp. 459–464.

Perera, M. D. A. E. S., 1982, Review of Techniques for Measuring Ventilation Rates in Multicelled Buildings, *EC Contractor's Meeting on Natural Ventilation, 1982, Energy Conservation in Buildings: Heating Ventilation and Insulation Bruxelles*, Reidl Publishing Company, Dordrecht.

Presser, K. H. and R. Becker, 1988, Mit Lachgas dem Luftstrom auf der Spur Luftstrommessung in Raumlufttechnischen Anlagen mit Hilfe der Spurgasmethode, *Heizung Luftung Haustechnik*, vol. 39, pp. 7–14.

Riffat, S. B. and S. F. Lee, 1990, Turbulent flow in a duct: Measurement by a tracer gas technique, *Building. Serv. Eng. Res. Technol.*, vol. 11, pp. 21–26.

Roulet, C.-A., 2004, *Santé et Qualité de l'Environnement Intérieur dans les Bâtiments*, PPUR, Lausanne.

Roulet, C.-A. and R. Compagnon, 1989, Multizone gas tracer infiltration measurement: Interpretation algorithms for non-isothermal cases, *Energy and Environment*, vol. 24, pp. 221–227.

Roulet, C.-A. and P. Cretton, 1992, Field comparison of age of air measurement techniques, *Roomvent 1992 conference*, Aalborg (DK), pp. 213–229.

Roulet, C.-A., and F. Foradini, 2002, Simple and cheap air change rate measurement using CO_2 concentration decays: *Int. J. of Ventilation*, vol. 1, p. 39.

Roulet, C.-A. and L. Vandaele, 1991, Airflow patterns within buildings: Measurement techniques, *AIVC Technical Note 34*, AIVC, Bracknell.

Roulet, C.-A. and M. S. Zuraimi, 2003, Applying tracer gas technique for measurements in air handling units with large recirculation ratio, *Healthy Buildings 2003 conference*, vol. 2, Singapore, pp. 536–541.

Roulet, C.-A., R. Compagnon and M. Jakob, 1991, A simple method using tracer gas to identify the main airflow and contaminant paths within a room, *Indoor Air*, vol. 3, pp. 311–322.

Roulet, C.-A., F. Foradini, P. Cretton, and M. Schoch, 1998, Measurements of ventilation efficiency in a retrofitted conference room.: *EPIC '98 conference*, Lyon, pp. 498–503.

Roulet, C.-A., F. Foradini and L. Deschamps, 1999, Measurement of air flow rates and ventilation efficiency in air handling units, *Indoor Air 1999 conference*, vol. 5, Edinburgh, pp. 1–6.

Roulet, C.-A., M.-C. Pibiri and R. Knutti, 2000, Measurement of VOC transfer in rotating heat exchangers, *Healthy Buildings 2000 conference*, vol. 2, Helsinki, pp. 221–226.

Roulet, C.-A., L. Deschamps, M.-C. Pibiri, and F. Foradini, 2000a, DAHU: Diagnosis of Air Handling Units, in H. B. Awbi, ed., *Air Distribution in Rooms – Ventilation for Health and Sustainable Development*, vol. 2, Elsevier, Amsterdam, pp. 861–866.

Roulet, C.-A., F. D. Heidt, F. Foradini and M. C. Pibiri, 2001, Real heat recovery with air handling units, *Energy and Buildings*, vol. 33, pp. 495–502.

Roulet, C.-A., M.-C. Pibiri, and F. Foradini, 2001a, Diagnostic des installations de ventilation – Méthodes et quelques résultats, *CIFQ 2001 – Vème Colloque Interuniversitaire Franco-Québecois – Thermique des Systèmes*, pp. 5–13.

Roulet, C.-A., F. Foradini, C. Cox, M. Maroni and E. D. O. Fernandes, 2005, Creating healthy and energy-efficient buildings: Lessons learned from the HOPE project, *Indoor Air conference 2005*, Beijing, Paper 1.6.44.

Ruud, S. and T. Carlsson, 1996, Transfer of pollutants in rotary air-to-air heat exchangers: A state of the art investigation, *Indoor Air 1996 conference*, Nagoya, pp. 977–982.

Ruysselvelt, P., 1987, Ventilation and heat recovery in superinsulated houses, *UK-ISES Conference Proceedings*, Hamburg, pp. 54–67.

Sandberg, M., 1984, The multi-chamber theory reconsidered from the viewpoint of air quality studies, *Building and Environment*, vol. 19, pp. 221–233.

Sandberg, M., and M. Sjöberg, 1983, The use of moments for assessing air quality in ventilated rooms, *Building and Environment*, vol. 18, pp. 181–197.

Sandberg, M. and C. Blomqvist, 1985, A quantitative estimate of the accuracy of tracer gas methods for the determination of the ventilation flow rate in buildings, *Building and Environment*, vol. 20, pp. 139–150.

Seibu Giken Co. Ltd., 1999, *Technical Information on Ion Power Total Heat Exchanger*, Seibu Giken, Fukuoka.

Sherman, M. H., 1990, Tracer gas techniques for measuring ventilation in a single zone, *Building and Environment*, vol. 25, pp. 365–374.

Sherman, M. H. and D. J. Dickerhoff, 1989, *A Multi-gas Tracer System for Multi-zone Air Flow Measurements*, ASHRAE/DOE/BTECC Symposium on Thermal Performance of External Envelopes of Buildings IV, Orlando, Florida.

Sherman, M. H., D. T. Grimsrud, P. E. Condon and B. V. Smith, 1980, Air infiltration measurement techniques, *First AIC Conference*, Windsor Great Park, Berkshire, UK, AIVC, pp. 9–44.

Silva, A. R. and C. F. Afonso, 2004, Tracer gas dispersion in ducts: Study of a new compact device using arrays of sonic micro jets, *Energy and Buildings*, vol. 36, pp. 1131–1138.

Sinden, F. W., 1978, Multi-chamber theory of air infiltration, *Building and Environment*, vol. 13, pp. 21–28.

Sutcliffe, H. C., 1990, A guide to air changes efficency, AIVC Technical Note 28, Bracknell, UK. Order at inive@bbri.be.

Tamura, G. T. and A. G. Wilson, 1966, Pressure differences for nine-storey building as a result of chimney effect and ventilation system operation, *ASHRAE Trans.*, vol. 72, pp. 180–189.

Tarantola, A., 1987, *Inverse Problem Theory: Method for Data Fitting and Model Parameter Estimation*, Elsevier, Amsterdam.

Valton, P., 1989, Renouvellement d'air dans les bâtiments, *PROMOCLIM E*, vol. 18, pp. 279–297.

Van der Maas, J., J. L. M. Hensen and A. Roos, 1994, Ventilation and energy flow through large vertical openings in buildings, *15th AIVC Conference*, Buxton, UK, vol. 1, pp. 289–302.

van der Wal, J. F., A. W. Hoogenvenn and L. V. Leeuwen, 1998, A quick screening method for sorption effects of volatile organic compounds on indoor materials, *Indoor Air*, vol. 8, pp. 103–112.

Wolkoff, P., P. A. Clausen, B. Jensen, G. D. Nielsen and C. K. Wilkins, 1997, Are we measuring the relevant indoor pollutants? *Indoor Air*, vol. 7, pp. 92–106.

Zwillinger, D., ed., 2003, *CRC Standard Mathematical Tables and Formulae*, CRC Press, Boca Raton, Florida.

Annex A

Unit Conversion Tables

Introduction

SI units are used throughout this book. Non-SI units are, however, of general use in air infiltration and ventilation, like the air change rate in l/hour or US units. To expedite the unit's translations, some tables are given below. Only physical quantities which are of general use in air infiltration and ventilation measurement techniques are listed.

The figures given in the tables are multiplying factors transforming values expressed in units of the first column into values expressed in the first row units. Example: 1 cm = 0.01 m.

Multiples and sub-multiples

	Multiples			Sub-multiples		
Prefix	Symbol	Factor	Prefix	Symbol	Factor
peta~	P	10^{15}	femto~	f	10^{-15}
tera~	T	10^{12}	pico~	p	10^{-12}
giga~	G	10^{9}	nano~	n	10^{-9}
mega~	M	10^{6}	micro~	μ	10^{-6}
kilo~	k	10^{3}	milli~	m	10^{-3}
hecto~	h	10^{2}	centi~	c	10^{-2}
deca~	da	10^{1}	deci~	d	10^{-1}

Length

Name	Symbol	m	cm	in	ft	yd
1 metre	m	1	100	39.37008	3.28084	1.093613
1 centimetre	cm	0.01	1	0.3937008	0.0328084	0.01093613
1 inch	in	0.0254	2.54	1	1/12	1/36
1 foot	ft	0.3048	30.48	12	1	1/3
1 yard	yd	0.9144	91.44	36	3	1

Area

Name	Symbol	m^2	cm^2	sq in	sq ft	sq yd
1 square metre	m^2	1	104	1550	10.7639	1.19599
1 square centimetre	cm^2	10^{-4}	1	0.3937008	0.0328084	0.01093613
1 square inch	sq in	$6.4516 \cdot 10^{-4}$	6.4516	1	1/144	1/1296
1 square foot	sq ft	0.092903	929.0304	144	1	1/9
1 square yard	sq yd	0.836127	8361.27	1296	9	1

Volume

Name	Symbol	m^3	l	$ml\ cm^3$	cu yd	cu ft	cu in
1 cubic metre	m^3	1	1000	10^6	1.30795	35.31464	5085.308
1 litre	l	0.001	1	1000	$1.308 \cdot 10^{-3}$	0.035315	5.085308
1 millilitre*	ml	0.000001	0.001	1	$1.308 \cdot 10^{-6}$	$35.32 \cdot 10^{-6}$	0.005085
1 cubic yard	cu yd	0.76455551	764.5555	764555.5	1	27	3888
1 cubic foot	cu ft	0.02831687	28.31687	28316.87	0.037037	1	144
1 cubic inch	cu in	$196.6 \cdot 10^{-6}$	0.196645	196.6449	$257.2 \cdot 10^{-6}$	$6.944 \cdot 10^{-3}$	1

Note: * the millilitre is equal to the cubic centimetre.

Mass

Name	Symbol	kg	g	lb	oz	gr
1 kilogram	kg	1	1000	2.204623	35.27396	15,432
1 gram	g	0.001	1	$2.205 \cdot 10^{-3}$	0.03527396	15.43
1 pound	lb	0.45359229	453.5923	1	16	700
1 ounce	oz	0.02834952	28.34953	0.0625	1	437
1 grain	gr	$64.79 \cdot 10^{-6}$	0.064799	$142.9 \cdot 10^{-6}$	$2.2857 \cdot 10^{-3}$	1

Time

Name	Symbol	s	min	h	d	yr
1 second	s	1	1/60	1/3600	1/86400	$31,688 \cdot 10^{-9}$
1 minute	min	60	1	1/60	1/1440	$1.90133 \cdot 10^{-6}$
1 hour	h	3600	60	1	1/24	$114.08 \cdot 10^{-6}$
1 day	d	86,400	1440	24	1	$2.73791 \cdot 10^{-3}$
1 year	yr	31,556,926	526103	8765	365.25	1

Pressure

Name	Symbol	Pa	mbar	mm H₂O	in H₂O	psi
I Pascal	Pa	I	0.01	0.102	0.004	$145.037 \cdot 10^{-6}$
I millibar	mbar	100	I	10.2	0.422	$14.5037 \cdot 10^{-3}$
I mm water column	mm H₂O	9.81	0.0981	I	0.0393	$1.42 \cdot 10^{-3}$
I inch water column	in H₂O	249	2.5	25.4	I	$36 \cdot 10^{-3}$
I pound per square inch	lb/in² or psi	6894.76	68.9476	703	27.7	I

Volume flow rate

Symbol	m³/s	l/min	m³/h	cu ft/s	cu ft/min	cu ft/h
m³/s	I	60,000	3600	35.3146	2118.878	127132.693
l/min	$16.667 \cdot 10^{-6}$	I	0.06	$588.58 \cdot 10^{-6}$	0.0353146	2.11887822
m³/h	$277.78 \cdot 10^{-6}$	16.666667	I	0.00980962	0.58857728	3
cu ft/s	0.02831687	1699.0122	101.9407335	I	60	3600
cu ft/min	$471.95 \cdot 10^{-6}$	28.316870	1.699012225	0.01666667	I	60
cu ft/h	$7.87 \cdot 10^{-6}$	0.47194784	0.028316870	$277.78 \cdot 10^{-6}$	0.01666667	I

Mass flow rate

Symbol	kg/s	kg/min	kg/h	lb/s	lb/min	lb/h
kg/s	I	60	3600	2.204623	132.27738	7,936.643
kg/min	0.01666667	I	60	0.03674372	2.204623	132.27738
kg/h	$277.78 \cdot 10^{-6}$	0.01666667	I	$612.4 \cdot 10^{-6}$	0.03674372	2.204623
lb/s	0.45359229	27.2155375	1632.932	I	60	3600
lb/min	0.00755987	0.45359229	27.21554	0.01666667	I	60
lb/h	$125.1 \cdot 10^{-6}$	0.00755987	0.4535923	$277.8 \cdot 10^{-6}$	0.01666667	I

Annex B

Glossary

Items in *italics* are additional entries in the glossary.

Age of the air (or age of a contaminant)
Average time period since the fresh air (or a contaminant) entered the room or the building. This age depends on the location in the building. The *room mean age of air* is the average of the age over the whole room.

Air change performance
Coefficient defined by ASHRAE, which is the double of the *air exchange efficiency*.

Air change rate (or *specific airflow rate*)
The ratio of the volumetric rate at which air enters (or leaves) an enclosed space divided by the volume of that space. Often this is expressed in air changes per hour. Its inverse is the *nominal time constant*.

Air exchange efficiency
Efficiency of the ventilation to change the air in a room. It is half the ratio of the *nominal time constant* and the *room mean age of air*.

Air exchange rate
General term relating to the rate of airflow between one space and another. This can be between various internal zones of a building or between the building and the atmosphere.

Air exfiltration
The uncontrolled leakage of air out of a building.

Airflow coefficient
Coefficient in the air *leakage characteristics*, which has the dimension of an airflow. This coefficient multiplies the *pressure differential* at a power exponent.

Airflow rate
The mass or volume of air moved in unit of time. (The transport may be within an enclosure or through an enclosing envelope.)

Air infiltration
The uncontrolled inward *air leakage* through cracks and interstices in any building element and around windows and doors of a building (i.e.,

adventitious openings), caused by pressure effects of the wind and/or the effect of differences in the indoor and outdoor air density.

Air infiltration characteristic

The relationship between the infiltration *airflow rate* into a building and the parameters that cause the movement.

Air leakage

Airflow rate through a component of the building envelope, or the building envelope itself, when a pressure difference is applied across the component.

Air leakage characteristic

An expression that describes the *air leakage* rate of a building or component. This may be:

* the air leakage flow rate at a reference pressure difference across the component or building envelope;
* the relationship between flow rate and the pressure difference across the building envelope or component;
* the *equivalent leakage area* at a reference pressure difference across the component or building envelope.

Airtightness

A general descriptive term for the *leakage characteristics* of a building.

Analyser

Instrument used to measure the *concentration* of a *tracer gas* or a *contaminant* in a sample of air.

Anemometer

Any instrument measuring the air speed or the air velocity.

Background concentration

Concentration of a gas in outdoor air.

Background leakage

Leakage of air through a building envelope that is not accounted for by obvious measurable gaps.

Balanced fan pressurization

Technique utilizing two or more *blower doors* to evaluate the leakage of individual internal partitions and external walls of multi-zone buildings. Technique involves using the fans to induce a zero pressure difference across certain building components, thus eliminating their leakage from the measurement.

Balanced ventilation

Ventilation systems in which fans both supply and extract air from the enclosed space, the supply and extract flow rates being equal.

Blower door (or fan door)

A device that fits into a doorway for supplying or extracting a measured flow rate of air to or from a building. It is normally used for testing for air leakage by *pressurization* or depressurization.

Building component
General term for any individual part of the building envelope. Usually applied to doors, windows and walls.

Building envelope
The total of the boundary surfaces of a building, through which heat (or air) is transferred between the internal spaces and the outside environment.

Calibration
Operation where the output of a measuring device is compared with reference standards, to accurately quantify the results provided by the measuring device.

Capacitance pressure transducer
A device with a metal diaphragm sensing element acting as one plate of a capacitor. When pressure is applied it moves with respect to a fixed plate, changing the thickness of the dielectric between. The resulting signal is monitored using a bridge circuit.

Cell
Volume in a *room* limited by a theoretical or physical surface, in which the physical quantities of interest can be considered as homogeneous. A room can be divided in several cells.

Chemical indicator tubes (or Dräger® tubes)
Glass tubes containing an adsorbing material that changes colour in the presence of a specific gas.

Compensated flowmeter
Airflow rate measuring instrument in which a fan compensates the pressure drop required by the measuring device.

Component leakage
The leakage of air through the building envelope or internal partitions, which is directly attributable to flow through cracks around doors, windows and other components.

Concentration
Ratio expressing the amount of a chemical component in a mixture. This ratio may be expressed in terms of mass, of volume or of number of molecules. In air, it can also be the ratio of the mass of component divided by the volume of air.

Condition number
Number expressing how much the errors in measured data are enlarged when transmitted, through the interpreting equations, to the final results.

Conductance
Generally, any path allowed for air between two *zones*. Also the ratio of the flow rate through a path to the *pressure differential* across that path.

Connected space
A space in a building adjacent to the measurement space with which significant exchange of air may take place, thus increasing the *effective volume* of the space.

Constant concentration technique

A method of measuring ventilation rate whereby an automated system injects *tracer gas* at the rate required to maintain the concentration of tracer gas at a fixed, predetermined level. The ventilation rate is proportional to the rate at which the tracer gas must be injected.

Constant injection rate technique

A method of measuring ventilation rate whereby tracer is emitted continuously at a uniform rate. The equilibrium concentration of *tracer gas* in air is measured.

Contaminant

An unwanted airborne constituent that may reduce the acceptability of the air quality.

Contaminant removal effectiveness

See *ventilation efficiency*.

Continuity equation (or *mass balance*)

Mathematical expression relating to the conservation of matter, an example of which is the equation equating the flow of *tracer gas* into a space with the flow of tracer gas out of a space. this particular equation is the basis for evaluating air exchange rates from tracer gas measurement.

Damper

Adjustable plate in a duct for controlling the flow rate.

Decay rate technique

A method for measuring ventilation rate whereby a quantity of *tracer gas* is released and the decrease in concentration measured as a function of time.

Deduction method

Multi-fan testing method in which the *pressure differential* between two *zones* of a building is changed step by step in order to obtain the *leakage characteristics* of building elements in these zones.

Density

Ratio of the mass of a quantity of matter to its volume.

Depressurization

Term used to describe *fan pressurization* when a static under-pressure is created within the building.

Differential pressure

See *pressure differential*.

Discharge coefficient

A dimensionless coefficient relating the mean flow rate through an opening to an area and the corresponding pressure difference across the opening.

Displacement flow (or *piston flow*)

With this type of flow, incoming outdoor air displaces internal air without mixing.

Distribution effectiveness

Ratio of the average *tracer gas* or *contaminant concentration* to the concentration that could be reached, at equilibrium, in the same *zone* or building with the same tracer or contaminant sources. Also the ratio of the contaminant or tracer *turnover time* to the *room mean age of air*. It is the inverse of the relative *contaminant removal effectiveness*.

Door panel

Panel adapted to a door or a window on which the *pressurization* fan is mounted.

Draught gauge

Inclined u-tube *manometer*.

Dräger® tubes

See *chemical indicator tubes*.

Effective volume

The volume of the interior building (or *room*) in which mixing occurs.

Efficiency of the ventilation system

Ratio of the fresh air provided by the ventilation system to an enclosure to the total amount of air entering the room, including *infiltration*.

Electron capture detector

An instrument, which uses a weak beta source to generate electrons in an ionization chamber, subjected to a pulsed voltage, thus generating a current. Electron-capturing material in the sample reduces the number of electrons in the chamber and thus the current. This reduction can be calibrated in terms of *tracer gas concentration*; hence the concentration of tracer gas in an air sample can be evaluated.

Envelope (of a building)

See *building envelope*.

Equivalent leakage area

The equivalent amount of orifice area that would pass the same quantity of air as would pass collectively through the building envelope at a specified reference pressure difference.

Experimental design

The way an experiment is planned, or, more precisely, a list of values of controlled parameters at which measurements should be performed to obtain the required results.

Extract ventilation

A *mechanical ventilation* system, in which air is extracted from a space or spaces, thus creating an internal negative pressure. Supply air is drawn through adventitious or intentional openings.

Fan pressurization

General term applied to any technique involving the production of a steady static *pressure differential* across a building envelope or component. Often referred to as *dc pressurization*.

Flame ionization detector

Detector used in conjunction with a *gas chromatograph*, in which the change in ionic current caused in a hydrogen–air flame by a tracer or contaminant is detected. This detector is sensitive to organic compounds.

Flow coefficient

In the power function approach this parameter is used in conjunction with the *flow exponent* to quantify flow through an opening.

Flow equation

Equation describing the *airflow rate* through a building (or component) in response to the pressure difference across the building (or component). These equations are usually *power law* or *quadratic law* in form.

Flow exponent

In the power function approach, this parameter characterizes the type of flow through a component ($n = 1$ represents laminar flow, $n = 0.5$ represents turbulent flow). For most flow paths, n takes a value between these extremes.

Fortuitous leakage

Uncontrolled air leakage through a building envelope due to the natural action of wind and temperature, i.e., *air infiltration*.

Gas chromatography

A process by which gases can be separated from one another. Used in this application to separate *tracer gases* from each other and from the constituents of air, thus allowing individual analyses to be performed.

Gasometer

Instruments to measure volumes of any gas.

Grab sampling method

Any *tracer gas* method where air/tracer samples are obtained from a building and analysed afterwards in a laboratory.

Guard zone technique

Dual fan *pressurization* technique used to measure the *leakage characteristics* of a building part. One fan is used to pressurize a guarding *zone*, surrounding the guarded zone in which the other fan just maintains a zero *pressure differential* between these zones. The measured building part is the only unguarded part.

Hot wire anemometer

Anemometer in which the temperature of a heated wire exposed to the wind determines the air velocity.

Indoor air pollution
Pollution occurring indoors from any source, i.e., from outside as well as inside the building.

Infrared gas analyser
Instrument used to determine *tracer gas concentrations* by determining the transmission of infrared radiation at an absorption frequency through a fixed path length.

Inter-zonal airflow
General term applied to the process of air exchange between internal *zones* of a building.

Leakage area
See *equivalent leakage area*.

Leakage characteristics
Equation relating the *airflow rate* through a leak and the *pressure differential* across this leak. This relation involves the *flow coefficient* and the *flow exponent*.

Leakage path
A route by which air enters or leaves the building or flows through a component.

Leakage site
A point on the outer or inner surfaces of a *building envelope* or an internal wall where a *leakage path* emerges.

Leeward
Downwind side of any object.

Manometer
A device for measuring pressure employing the principle of displacement of liquid levels in a liquid-filled u-tube. The limbs of the 'u' may be vertical, inclined (*draught gauge*) or curved.

Mass balance
See *continuity equation*.

Mass flow controller
Device controlling the flow rate of a gas by means of a valve controlled according the measurement of the mass flow rate.

Mass spectrometry
Technique allowing the quantitative measurement of amounts of different gases, based on the separation of the ionized gas molecules according their mass-to-charge ratio.

Mechanical ventilation
Ventilation by means of one or more fans.

Mixing
The degree of uniformity of distribution of outdoor air or foreign material in a building.

Mixing fan
Small electric fan used to aid the mixing of room air and *tracer gas* before and/or during a measurement.

Multiple tracer gas technique
General term applied to measurement methods using two or more *tracer gases*. These methods are often used to evaluate inter-zonal airflows.

Multi-zone
A building or part of a building comprising a number of *zones* or *cells*.

Natural ventilation
Ventilation using only purpose-provided openings and the natural motive forces of wind and temperature difference.

Nominal time constant
The ratio of the volume of an enclosed space divided by the volumetric rate at which air enters (or leaves) that space. Its inverse is the *air change rate*.

Normalized leakage area
Equivalent leakage area expressed per unit building envelope area.

Orifice plate
A device for assessing gas flow by measuring the pressure drop across an orifice in the flow line.

Outdoor air
Air from free atmosphere that is generally assumed to be sufficiently uncontaminated to be used for ventilation.

Passive sampling
Method of sampling *tracer gas* in a building by the process of passive diffusion.

Passive tracer source
Small container injecting continuously a small flow of tracer (mostly *PFT tracers*) by passive diffusion through its cover cap.

Perfluorocarbon tracers (or PFT)
Tracer gases composed of a family of perfluoroalkylcycloalkanes, i.e., cyclic organic compounds in which the hydrogen atoms are all replaced by fluorine atoms. These tracers can be analysed in trace amount because the background concentration is low and the *electron capture detector* is very sensitive to them.

Photo-acoustic detector
Tracer gas analyser in which the alternate expansion and contraction of the gas sample irradiated with a chopped beam of convenient wavelength is detected with a microphone.

Piston-type ventilation
See *displacement flow*.

Pitot tube
Anemometer measuring the difference between the pressure in a tube facing the flow, in which the flow is stopped, and the pressure along a side of the tube.

Pollutant removal effectiveness
See *ventilation efficiency*.

Pollution migration
Descriptive term for the movement of indoor air pollutants throughout a building.

Pollution source
Any object, usually within a building, that produces a substance that will contaminate the internal environment.

Power law
Flow equation in which the *airflow rate* through the *building envelope* is proportional to a power of the *pressure differential*.

Ppm
Unit for expressing volume concentration, which is a part per million (10^{-6}) or a cubic centimetre per cubic metre.

Pressurization
Airtightness measuring technique using a fan to pressurize the measured volume at a constant pressure. See also *fan pressurization*.

Pressure differential
Usual term for the difference in pressure across a *building envelope* or component, whether caused by natural or artificial means.

Pressure tap
Point at which pressure is measured.

Pulse injection technique
Tracer gas measuring technique in which the tracer is injected in a short pulse.

Purging flow rate
Part of the *airflow rate*, which effectively removes the *contaminants* out of the location of interest. It is the product of the airflow rate and the *ventilation efficiency*.

Purpose-provided openings
Openings in the *building envelope* for the specific purpose of supplying or extracting ventilation air.

Quadratic law
Flow equation in which the *pressure differential* is related to the *airflow rate* by a quadratic polynomial.

Reductive sealing method
A method of determining the leakage of specific building components by pressurizing the building and recording the leakage changes as components are sealed successively. When all the major outlets and component cracks are sealed, the remainder is the *background leakage*.

Relative contaminant removal effectiveness
Ratio of the concentration that could be reached, at equilibrium, in the same zone or building with the same tracer or contaminant sources, to the average tracer or *contaminant concentration*. Also the ratio of the *room mean age of air* to the contaminant or tracer *turnover time*. It is the inverse of the *distribution effectiveness*.

Residence time
See *age of the air*.

Residual gas analyser
See *mass spectrometry*.

Retrofit
The process of reducing energy loss in a building by physical means, for example, reducing excess air infiltration by obstructing flow through cracks and openings.

Reynolds number
Ratio of the inertial force to the friction force. It is also the ratio of the velocity of a fluid to its dynamic viscosity, multiplied by a typical dimension, for example, the duct diameter.

Room
Volume of a building limited by building elements. In ventilation technique, this concept keeps its usual meaning. A room may be divided in several *cells* and several rooms may be combined in a *zone*.

Room mean age of air
Average of the *mean age of air* over the whole *room*.

Sample container
Container used to obtain a sample of air/tracer mixture from a measured building. The sample is usually returned to a laboratory for analysis.

Short-circuiting
A direct flow path between an air supply point and an air extract point, i.e., air flows along the shortest path, without mixing.

Single tracer gas technique
General term applied to any method using only one *tracer gas*. These methods are usually used to evaluate *air change rate*.

Single zone
Any case where a building or part of a building is considered to be a single well-mixed space.

Site analysis
Applied to any *tracer gas* measurement technique where tracer gas *concentrations* and *air exchange rates* are determined directly at the measurement building.

Smoke leak visualization
A method of detecting leaks in the building fabric by pressurizing the building and using smoke to trace the paths followed by the leaking air.

Specific airflow rate (or *air change rate*)
The ratio of the volumetric rate at which air enters (or leaves) an enclosed space divided by the volume of that space. Its inverse is the *nominal time constant*.

Specific leakage area
Equivalent leakage area expressed per unit floor area.

Stack effect
Pressure differential across a *building envelope* caused by differences in the density of the air due to an indoor–outdoor temperature difference.

Step injection technique
Tracer gas measurement technique in which the tracer is injected at constant rate, starting from a given time.

Supply ventilation
A system in which air is supplied to a space(s) so creating an internal positive pressure. Air leaves the building through adventitious or purpose-provided openings.

Tachometer
Instrument for measuring velocity or speed of rotation, used to evaluate the speed of fans, this in turn is used to calibrate the fan in terms of airflow. Often used in *blower doors*.

Thermography
The process of converting the heat emitted from an object into visible pictures. Used to indicate and represent the temperature distribution over part of a building envelope.

Tracer gas
A gas used at low concentration, together with an analyser, to determine *airflow rates* or other related quantities.

Tracer gas analyser
Any instrument used to evaluate the concentration of *tracer gas* in a sample of air.

Tracer gas injection
Any process by which *tracer gas* is released into a space.

Tracer gas sampling
Any process by which *tracer gas* or air containing tracer gas is sampled for analysis.

Turnover time of a contaminant

Ratio of the mass of *contaminant* contained in an enclosure to the mass flow rate of the contaminant source in this enclosure.

Ventilation

The process of supplying and removing air by natural or mechanical means to and from any space.

Ventilation efficiency

An expression describing the ability of a mechanical (or natural) ventilation system to distribute the outdoor air in the ventilated space.

Ventilation energy

Energy loss from a building due to ventilation.

Venturi tube

Duct with a restricted section, which allows the measurement of the flow rate through the pressure differential between the restricted and the normal section.

Windward

Upwind side of any object.

Zone

Part of a building, which is considered as a single volume for the experiment performed, or the physical quantity of interest. A zone may contain several *rooms*.

Index

Printed and bound by CPI Group (UK) Ltd, Croydon, CR0 4YY

22/10/2024

01777621-0008